ISBN 978-0-265-94585-8
PIBN 10913750

W. B. No. 208.

U. S. DEPARTMENT OF AGRICULTURE,
U. S. WEATHER BUREAU.
BULLETIN F.

VERTICAL GRADIENTS

? 4 1.

OF

TEMPERATURE, HUMIDITY, AND WIND DIRECTION.

A PRELIMINARY REPORT ON THE KITE OBSERVATIONS OF 1898.

PREPARED UNDER DIRECTION OF WILLIS L. MOORE, CHIEF U. S. WEATHER BUREAU,

BY

H. C. FRANKENFIELD, Forecast Official.

————•———

WASHINGTON:
GOVERNMENT PRINTING OFFICE.
1899.

LETTER OF TRANSMITTAL.

U. S. Department of Agriculture,
Weather Bureau, Office of the Chief,
Washington, D. C.

Hon. James Wilson,
Secretary of Agriculture.

Sir: I have the honor to transmit for publication, herewith, a paper on Vertical Gradients of Temperature, Humidity, and Wind Direction, being a preliminary report on the kite observations of the year 1898, by Dr. H. C. Frankenfield, forecast official.

Dr. Frankenfield states that in the preparation of this paper he has freely consulted with Professors Abbe, Bigelow, and Marvin, of this Bureau. An introductory chapter on the Kite Meteorograph, Construction, and Operation is added by Prof. C. F. Marvin.

This report discusses probably the largest amount of free-air meteorological observations ever taken within a like space of time over so great an area, and I believe will be found to have added something to the general knowledge of temperature and hygrometric conditions of the lower levels of the air.

Very respectfully,

Willis L. Moore,
Chief United States Weather Bureau.

Approved:
James Wilson,
Secretary of Agriculture.

CONTENTS.

KITE METEOROGRAPH, CONSTRUCTION, AND OPERATION.

By Prof. C. F. MARVIN.

In designing the kites, meteorograph, and aerial apparatus employed in making the observations discussed in the accompanying paper, it was necessary to satisfy a number of conditions and limitations, as, for example, portability, simplicity, and strength of construction, especially as regards the kite; uniformity and interchangeability not only of parts of the apparatus, but of parts of the same devices, so that repairs and renewals could be made as easily as possible. The corps of observers, moreover, when they began, were without previous experience in kite flying. The entire class, however, reported at Washington before the individual members were assigned to station, and was put through a preliminary drill and course of instruction in the practical work. In the short period of ten to twenty days devoted to this instruction (including the time lost during unfavorable weather) it was possible for the observers to gather only a general knowledge of the apparatus and methods of work.

The kite.—In conformity with the foregoing limitations, it was considered best to employ in each ascension only one kite, rather than several in tandem, which latter are very troublesome to set flying in light and fitful winds and are not required; in fact, are less efficient than a single kite in favorable winds.

Nearly all observations were made with a medium size kite containing 68 square feet of supporting surface. At some stations a smaller and a slightly larger kite were also sometimes used, according to the strength of the wind. The smaller size contained 45 square feet and the larger 72 square feet. Fig. 1 shows the kite with the meteorograph in place. The dimensions of the medium size kite are as follows:

Transverse width of kite .... 6 feet 6 inches.
Length, over extreme edges, fore and aft[1] .... 6 feet 2¼ inches.
Distance between top and bottom supporting surfaces .... 2 feet 8 inches.
Width of cloth bands .... 2 feet.

These kites are framed and constructed in the most rigid manner possible. The six longitudinal sticks running fore and aft are attached by means of small machine screw bolts to the rectangular frames forming the rigid edges to the cloth bands and are detachable; thus permitting the kite to be collapsed, as shown in fig. 2.

The best cloth material for kites seems to be Lonsdale cambric, which is light, strong, and closely woven. A black, or dark-colored cloth is more visible on many occasions, and the rear cell was covered with black nainsook on this account. This cloth is not as strong as the cambric, but, owing to the circumstance that the pressure per unit area on kites of this type is very much greater in the forward than in the rear cell, the above disposition of the relatively strong and weak material is wholly justified. This same circumstance explains why the front cell is made with three supporting surfaces, as compared with two in the rear cell; that is to say, this increase of surface (at slight increase of weight) increases the lifting efficiency of the kite.

---

[1] The length, fore and aft, was increased to 6 feet 8¼ inches, in later forms.

The flying line is attached to the front edge of the forward cell at the middle; the ordinary arrangement of this connection being more fully shown in fig. 3.

Bridle and safety line.—The normal bridle, including the safety line for the kite, is shown in fig. 3. The cord (No. 32 Italian blocking cord) at A is passed twice around the stick, after the fashion of a "clove hitch;" the free portions of the line, one of which is rather short, are firmly bound together where they emerge from the clove hitch by a serving of waxed "gilling" thread. The long end is passed through one of the metal safety-line eyes, B, and the two ends tied together by means of bowline knots, as at b. A similar, but longer, piece of cord is secured at C by a clove hitch and the free ends attached to each other by bowline knots after the long end has been passed through the eye B'. When B' is held so that the line AB' is taut and at an angle of 90° to the stick, the line $C'B$ should be just a little taut. The main line is attached to the bridle at B'.

Safety line.—The eyes, BB', are connected by what is termed a safety line, S, which is simply a piece of steel wire, the size of which is so chosen that its ultimate strength is within a safe working strain for the kite and flying line. In normal flight all the strain produced by the kite upon the line is transmitted through the safety line S. If in any case the conditions give rise to a greater strain than the ultimate strength of the safety line (and therefore dangerous to either the kite or the line, or both), it will be broken. In this case the kite will thereafter fly from the point C, which is several inches in advance of A. Such a change in the bridle causes a diminished pull by the kite, other things being the same.

The safety lines generally supplied have a tensile strength of about 85 pounds.

The reel.—The management of large kites in flight requires a substantial and convenient form of reel of the character indicated in fig. 4. The top portion of the carriage revolves upon the table below on bearings resembling the so-called "fifth wheel" of wagons. The drum revolves easily in metal bearings and is fitted with dials at the axle indicating the number of revolutions. At the start the dials stand at zero and count off revolutions as the wire unwinds.

Strap brake.—The lever seen at the right in fig. 4 operates a powerful strap-iron friction brake acting on the rim of the drum and controls in the easiest and most complete manner the unwinding of the wire or the stoppage of the reel under all circumstances.

A matter of great importance in the design of the winding drum of the reel is to secure sufficient strength in the rim to withstand the enormous cumulative pressure exerted by a large amount of wire wound in under great tension. A single turn of wire around the drum, under a uniform strain of 50 pounds, for example, tends to produce a compressive stress of 50 pounds at every point around the rim. The next turn, at the same tension, adds 50 pounds to the preceding stress, and so on. Two thousand turns at this rate will, therefore, produce a pressure of 100,000 pounds, or 500 tons. The heavy rim of the cast-iron drum, shown in fig. 4, is calculated to safely resist a crushing pressure of fully 1,000 tons. In actual practice the crushing pressure is not quite so great as that calculated by the process indicated above, because the material of the reel yields a little as the pressure increases, and this lessens the tension on the turns of wire already wound on the drum. The side flanges of the drum must also be very strong, as the wire crowds sidewise against these with great force. It is best on this account not to wind the wire on in smooth and even layers, but rather to crisscross the turns of wire slightly, but in a regular manner. Wound in this way, the wire tends to support itself, even without side flanges; at any rate, the lateral pressure is greatly reduced, and, moreover, the outside turns of wire are not able to squeeze down through what is already wound on the reel, as they tend to do when the wire is wound in an even manner, like thread on a spool.

When flying at an elevation of from 5,000 to 7,000 feet, one of the Weather Bureau kites, supporting its instrument, will pull from 60 to 80 pounds, if not more, and from 8,000 to 10,000 feet of wire will be out. To wind all this wire in under such conditions is really a very laborious operation, and generally requires two men at pretty hard work for from a half to three-quarters of an hour or more.

As sent out to stations the hand reels contained from 2,600 to 3,000 turns of tempered steel music wire, 0.028 of an inch in diameter. The normal tensile strength of this wire was about 200 pounds.

Length of wire.—As the original supply of wire was wound upon each reel, record was kept of the total number of turns and a table computed, giving the number of turns corresponding to given lengths of wire in units of 500 feet. Due account is taken in these tables of the gradual diminution in the length of each turn as more and more wire is unwound. The coefficient of diminution was determined from several sets of readings of the revolutions of a measuring wheel around which the wire passed as it was being wound on a reel. Simultaneous readings of the dial on the reel were also made. The measuring wheel was accurately 3 feet in circumference and the dial indicated feet.

Electrical connections.—The wire line employed in flying kites becomes electrified more or less at all times, often highly so. For comfort of the operators, as well as safety, this charge must be conveyed to earth, and for this purpose each reel stand is provided with an electric ground-connection and switch. The 15-inch cranks by which the reel is revolved are made of wood for the sake of insulation.

Radius rod and arc.—It is important to know the inclination of the wire at the reel in order to make a proper allowance for the sag of the wire. This is accomplished by means of the radius rod and the graduated arc, fig. 4. The radius rod is clasped loosely upon the axle of the reel on either side of the drum, and the arc hung over the shaft on a pair of antifriction wheels which run in a groove turned in the shaft. A weighted rod below the arc causes it to maintain a vertical position at all times, thus insuring correct angles. In use, the radius rod is made to rest against the wire, the angular inclination of which to the horizontal is then shown by the reading upon the graduated arc. This angle is subject to a small and variable inaccuracy, due to slight alterations in the radial distance of the wire at the point it leaves the drum, according as more or less of the wire is unwound.

Dynamometer.—The tension upon the line at the reel at any time is determined by means of the dynamometer permanently attached to one of the crank handles, as seen in fig. 4. This consists of a short, stiff, steel spring, firmly fastened to the outer end of the handle at the back. The short end of a multiplying lever connects with the spring, while the long end serves as an index and traverses the graduated arc shown on the crank handle near the axis. The reading on this graduated arc indicates the pull in pounds on the wire. Here, likewise, a small and variable error is introduced because of the variations in the diameter of the drum with different amounts of wire out. This is not important.

End of wire.—The outer end of the wire terminates in a small brass eye. To facilitate connecting the wire to the kite, a piece of No. 32 blocking cord about 8 feet long is fastened to the eye—that is to say, the cord is simply passed through the eye and a bowline knot tied on one end.

Reel box, cover, and lock.—When not in use it is designed simply to inclose the reel within the box by means of a suitable cover. The crank handles are unshipped and placed inside the box. The cover is secured and locked by means of an eyebar, which passes between the spokes of the drum and is fastened by a padlock on one end.

The meteorograph.—The instrument sent up with the kite to secure the automatic record of the conditions of the air is seen in fig. 1 as it appears attached to the kite and inclosed within its light aluminum case. The mechanisms inside the case are shown in fig. 5, and are designed to record wind velocity, temperature, pressure, and humidity of the air. Records of the velocity of the wind were, however, made experimentally only at Washington, and were not included in the official observations.

Special consideration was given in designing the meteorograph to secure a proper exposure of the hair hygrometer and the thermograph bulb. The former consists of a strand of prepared hairs stretched back and forth in a double strand and nearly from end to end inside the long tube seen at the top portion of the instrument, as shown in fig. 5. The direct elongation and contraction of these hairs is communicated to the recording pen at the extreme right side in fig. 5. The thermograph bulb is also placed within the long tube and occupies nearly a middle position.

When attached to the kite, as shown in fig. 1, the meteorograph is so placed that the wind blows with full force directly through the tube containing the thermometer bulb and hygrometer, thus affording thorough and complete ventilation, radiation being at the same time effectually cut off. Even though the metallic case of the instrument becomes heated on exposure to sunshine, yet the metal tube itself is not only shaded and exposed to a strong current of wind, but is everywhere separated from contact with the metallic case by vulcanite rings at the ends and longitudinal ivory strips on the sides. Indeed, celluloid strips are interposed between the thermometer bulbs and the metal of the inclosing tube, thereby still further insulating the thermometer bulbs. These bulbs consist of a pair of tempered-steel bourdon pressure tubes, forming a curl of about seven-eighths of a complete circle about 1¼ inches in diameter. The major and minor axes of the elliptical cross section of the tubes measure approximately 0.5 and 0.1 of an inch, respectively. The tubes are filled with pure alcohol under pressure, and are set in tandem and edgewise to the current of wind through the tube. Thus arranged, they constitute highly sensitive and relatively powerful thermometric bulbs. The recording pen traverses a scale of 22 degrees to the inch, each instrument being adjusted to this scale by tests at different temperatures, the air being driven through the tube by means of an electric fan. The record sheet provides a range of 45 degrees possible change of temperature in any ascension, the initial setting of the thermograph pen being effected by a suitable adjusting screw.

The mechanism of the aneroid barometer is readily understood from fig. 5. The novel feature in this part of the meteorograph consists in the use of tempered and highly elastic steel corrugated disks, instead of brass or German silver ordinarily used. Numerous tests showed the steel vacuum chambers to be superior to others. Nevertheless the aneroid principle has not thus far proved to give pressure records characterized by that high degree of precision required in meteorological work.

The pressure scale on the record sheet embraces a range of 9 inches, viz, 21 to 30 inches air pressure. The subdivisions of the sheet are 22 spaces per inch, each space representing 0.2 of an inch barometric pressure.

It will be noticed these spaces are the same as the degree spaces on the temperature record. This arrangement was chosen so that in extreme cases the temperature record might overlap on the space normally provided for pressure, and vice versa, and still be properly scaled by the rulings on the sheet. Furthermore, in designing the instrument the scales were so chosen that in an ascension under average atmospheric conditions the pressure and temperature curves would have about the same actual amplitude. Hence the change of temperature in a given elevation is measured with the same accuracy as the corresponding change in pressure, at least so far as the traces themselves are concerned. A greater precision of pressure measurement is desirable, but is scarcely justified by the inherent defects of the aneroid principle of measurement.

Each recording pen is adjustable, and at or before the time of ascension the pens are set as nearly as may be to indicate correctly the atmospheric conditions as shown by readings of the sling psychrometer and mercurial barometer. After these settings are made and just before ascensions, the meteorograph having been exposed for some time in the wind, readings of the psychrometer and barometer are again made and the outstanding differences of the indications of the meteorograph noted and used as corrections to the automatic records, similar corrections being noted at the close of the ascension.

Elevation of the kite.—The vertical height of the kite and hence the meteorograph above ground was determined from readings of the angular elevation of the kite, the length of line out, and its inclination at the reel. The angular elevation was measured to the nearest half degree by means of the nephoscope, shown in fig. 6. The position of the sighting staff also indicated the angular azimuth of the kite.

The approximate elevation of the kite was taken out from a table giving values of the expression $h = l \sin \varphi$. A percentage correction was then applied, depending upon the amount of sag in the wire as shown by its observed inclination at the reel.

Table I gives the percentage of slack in the wire, as deduced from the equation of the catenary for such conditions as commonly occur.

Table II is a copy of one of the cards furnished with reel No. 15, giving the number of turns of the reel corresponding to the lengths of wire out in 500-foot intervals.

As a rule, observations were made with whole units of 500 feet of wire out, so that interpolation was not necessary. The length of wire in one turn was, however, accurately tabulated, and facilitated accurate interpolation.

TABLE I.—ANGULAR ELEVATION AND PERCENTAGE OF SLACK.

Angular elevation of kite.	Inclination of wire at reel.																Angular elevation of kite.
	0°	5°	10°	12°	14°	16°	18°	20°	22°	24°	26°	28°	30°	35°	40°	45°	
	Percentage of slack.																
24	2.4	1.6	0.9	0.7	0.5												24
26	2.7	1.8	1.1	0.9	0.7	0.5											26
28	3.1	2.2	1.4	1.1	0.9	0.7	0.5	0.3									28
30	3.4	2.5	1.6	1.3	1.1	0.9	0.7	0.5	0.3								30
32	3.7	2.8	1.9	1.6	1.3	1.1	0.8	0.6	0.4	0.3							32
34	4.1	3.1	2.2	1.9	1.6	1.3	1.0	0.8	0.6	0.4	0.3	0.1	0.0				34
36	4.4	3.4	2.5	2.2	1.8	1.5	1.2	1.0	0.8	0.6	0.4	0.2	0.1				36
38	4.6	3.6	2.7	2.4	2.1	1.8	1.5	1.2	1.0	0.8	0.6	0.4	0.3				38
40	4.9	3.9	3.0	2.7	2.4	2.1	1.8	1.5	1.2	1.0	0.8	0.6	0.4	0.2			40
42	5.1	4.2	3.3	3.0	2.6	2.3	2.0	1.7	1.4	1.2	1.0	0.8	0.6	0.3			42
44	5.3	4.4	3.5	3.2	2.8	2.5	2.2	1.9	1.6	1.4	1.2	1.0	0.8	0.4	0.1		44
46	5.5	4.6	3.7	3.4	3.1	2.8	2.5	2.2	1.9	1.6	1.4	1.1	0.9	0.5	0.2		46
48	5.7	4.8	3.9	3.6	3.3	3.0	2.7	2.4	2.1	1.8	1.6	1.3	1.1	0.6	0.3	0.0	48
50	5.9	5.0	4.1	3.8	3.5	3.2	2.9	2.6	2.3	2.0	1.8	1.5	1.3	0.8	0.4	0.1	50
52			4.3	4.0	3.7	3.4	3.1	2.8	2.5	2.2	2.0	1.7	1.5	1.0	0.5	0.2	52
54								2.9	2.6	2.4	2.1	1.9	1.7	1.1	0.7	0.3	54
56													1.8	1.3	0.8	0.4	56

TABLE II.—SHOWING LENGTH OF WIRE UNWOUND FROM KITE REEL.

REEL No. 15.

Wire out.	Dial reading.	Length of one turn.
Feet.		Feet.
0	0	5.03
500	100	5.02
1,000	199	5.01
1,500	299	5.00
2,000	400	4.98
2,500	500	4.97
3,000	601	4.96
3,500	702	4.95
4,000	803	4.94
4,500	904	4.93
5,000	1,005	4.92
5,500	1,107	4.91
6,000	1,209	4.90
6,500	1,311	4.89
7,000	1,413	4.88
7,500	1,516	4.87
8,000	1,619	4.86
8,500	1,722	4.85
9,000	1,825	4.84
9,500	1,929	4.83
10,000	2,032	4.81
10,500	2,136	4.80
11,000	2,240	4.79
11,500	2,345	4.78
12,000	2,449	4.77
12,500	2,554	4.76
13,000	2,660	4.75
13,500	2,765	4.74
14,000	2,871	4.73
14,357	2,946	4.72

INSTRUCTIONS FOR USE OF TABLE SHOWING LENGTH OF WIRE UNWOUND FROM KITE REEL.

The table has been arranged to show dial readings corresponding to even lengths of 500 feet.

If the length of wire out is desired for a dial reading intermediate to those given in the table it can be determined by multiplying the difference between the *observed dial reading* and the nearest *tabulated dial reading* by the corresponding *length of one turn* of wire as given in the third column of table. The product should be added to or subtracted from the length in the first column opposite the tabulated dial reading used, according to whether the observed dial reading is greater or less than the tabulated dial reading. Example:

The observed dial reading is..........................2,000
The nearest tabulated dial reading is 2,032
The difference is 32
The corresponding length of one turn is...............4.82

$$4.82 \times 32 = 154$$
$$10,000 - 154 = 9,846$$

which is the length of wire out corresponding to the observed dial reading.

WILLIS L. MOORE,
Chief of Bureau.

For further particulars concerning the details of the aerial apparatus employed in the Weather Bureau work and data concerning the efficiency of kites, the reader is referred to the Monthly Weather Review, Vol. XXIV, 1896, p. 113; Vol. XXV, 1897, p. 136; Yearbook of the Department of Agriculture for 1898, p. 201, and "Instructions for Aerial Observers," 1898, W. B. No. 166.

VERTICAL GRADIENTS OF TEMPERATURE, HUMIDITY, AND WIND DIRECTION.

A PRELIMINARY REPORT ON THE KITE OBSERVATIONS OF 1898.

INTRODUCTORY.

Kite ascensions were commenced at seventeen stations during the latter portion of the month of April and were continued into the early days of November, 1898, but in the preparation of this report only those from May to October, inclusive, have been considered.

Ascensions were made on every day when it was possible to do so. For obvious reasons the kites could not be flown in rainy or extremely threatening weather, and a considerable number of days was lost on this account. The principal reason, however, for failures to obtain satisfactory ascensions daily was the absence of a sufficient wind velocity to sustain the kites. There were made 44 per cent of the total number of ascensions which would have been possible had the wind and weather conditions been favorable. The percentage of ascensions at the different stations varied from 75 at Dodge City, Kans., where the wind velocity was greatest, to 12 at Knoxville, Tenn., where it was least. The largest number of ascensions was obtained in the country west of the Mississippi River, and the least in the central river valleys.

The temperature conditions at all elevations and under varying conditions of weather and time have been computed in terms of the gradient in degrees Fahrenheit for each 1,000 feet, and in the increase of elevation necessary to cause a fall of 1 degree in the temperature. The mean results, however, are given only in degrees per thousand feet.

Some attention has also been given to the questions of wind directions, relative humidity, and vapor pressure.

The tables and plates accompanying will, it is believed, furnish all the tabular matter that may be desired.

SUMMARY OF OBSERVATIONS.

Stations.	Eleva-tion above sea level.	Number of months observa-tions.	Num-ber of ascen-sions.	Number of observations at—									Total.
				1,000 feet.	1,500 feet.	2,000 feet.	3,000 feet.	4,000 feet.	5,000 feet.	6,000 feet.	7,000 feet.	8,000 feet.	
	Feet.												
Washington, D. C.....	115	5.0	87	51	52	53	28	19	9	7	3	1	223
Cairo, Ill	315	5.5	39	6	33	31	23	19	4	.....	.....	.....	116
Cincinnati, Ohio......	940	5.0	38	3	7	24	19	16	9	3	.....	.....	81
Fort Smith, Ark	1 527	5.0	19	12	18	18	8	2	.....	.....	.....	.....	58
Knoxville, Tenn	990	5.0	19	8	8	8	8	2	.....	.....	.....	.....	34
Memphis, Tenn	319	4.3	37	16	22	27	26	21	5	.....	.....	.....	117
Springfield, Ill	684	5.0	46	15	52	53	28	15	9	2	.....	.....	174
Cleveland, Ohio	705	5.0	93	29	62	55	48	21	11	1	.....	.....	227
Duluth, Minn	1,197	5.5	96	61	71	83	60	32	13	5	.....	.....	325
Lansing, Mich........	869	5.0	58	23	23	22	34	22	10	.....	.....	.....	134
Sault Ste. Marie, Mich	722	5.3	74	7	42	46	39	31	15	.....	.....	.....	180
Dodge City, Kans.....	2,473	6.0	138	119	123	118	106	66	28	10	2	1	573
Dubuque, Iowa........	894	5.7	65	19	32	40	35	15	7	.....	.....	.....	148
North Platte, Nebr ...	2,811	6.0	132	70	162	133	102	44	13	1	.....	.....	525
Omaha, Nebr	1,241	4.5	61	0	29	50	45	34	19	6	2	.....	185
Pierre, S. Dak	1,595	5.5	134	96	105	91	65	38	19	2	.....	.....	416
Topeka, Kans	972	6.0	81	68	65	76	72	26	11	1	.....	.....	319
Total	.....	.....	1,217	603	906	928	746	423	182	38	7	2	3,835

1 516 after July 31.

14

TABLE FOR THE CONVERSION OF RATES INTO GRADIENTS.

Rate.	Gradient.	Rate.	Gradient.	Rate.	Gradient.	Rate.	Gradient.	Rate.	Gradient.
Feet.	°	Feet.	°	Feet.	°	Feet.	°	Feet.	°
100	10.00	176	5.68	252	3.97	328	3.05	404	2.48
101	9.90	177	5.65	253	3.95	329	3.04	405	2.47
102	9.80	178	5.62	254	3.94	330	3.03	406	2.46
103	9.71	179	5.59	255	3.92	331	3.02	407	2.46
104	9.62	180	5.56	256	3.91	332	3.01	408	2.45
105	9.52	181	5.52	257	3.89	333	3.00	409	2.44
106	9.43	182	5.49	258	3.88	334	2.99	410	2.44
107	9.35	183	5.46	259	3.86	335	2.99	411	2.43
108	9.26	184	5.43	260	3.85	336	2.98	412	2.43
109	9.17	185	5.41	261	3.83	337	2.97	413	2.42
110	9.09	186	5.38	262	3.82	338	2.96	414	2.42
111	9.01	187	5.35	263	3.80	339	2.95	415	2.41
112	8.93	188	5.32	264	3.79	340	2.94	416	2.40
113	8.85	189	5.29	265	3.77	341	2.93	417	2.40
114	8.77	190	5.26	266	3.76	342	2.92	418	2.39
115	8.70	191	5.24	267	3.75	343	2.92	419	2.39
116	8.62	192	5.21	268	3.73	344	2.91	420	2.38
117	8.55	193	5.18	269	3.72	345	2.90	421	2.38
118	8.47	194	5.15	270	3.70	346	2.89	422	2.37
119	8.40	195	5.13	271	3.69	347	2.88	423	2.36
120	8.33	196	5.10	272	3.68	348	2.87	424	2.36
121	8.26	197	5.08	273	3.66	349	2.87	425	2.35
122	8.20	198	5.05	274	3.65	350	2.86	426	2.35
123	8.13	199	5.03	275	3.64	351	2.85	427	2.34
124	8.06	200	5.00	276	3.62	352	2.84	428	2.34
125	8.00	201	4.98	277	3.61	353	2.83	429	2.33
126	7.94	202	4.95	278	3.60	354	2.82	430	2.33
127	7.87	203	4.93	279	3.58	355	2.82	431	2.32
128	7.81	204	4.90	280	3.57	356	2.81	432	2.31
129	7.75	205	4.88	281	3.56	357	2.80	433	2.31
130	7.69	206	4.85	282	3.55	358	2.79	434	2.30
131	7.63	207	4.83	283	3.53	359	2.79	435	2.30
132	7.58	208	4.81	284	3.52	360	2.78	436	2.29
133	7.52	209	4.78	285	3.51	361	2.77	437	2.29
134	7.46	210	4.76	286	3.50	362	2.76	438	2.28
135	7.41	211	4.74	287	3.48	363	2.75	439	2.28
136	7.35	212	4.72	288	3.47	364	2.75	440	2.27
137	7.30	213	4.69	289	3.46	365	2.74	441	2.27
138	7.25	214	4.67	290	3.45	366	2.73	442	2.26
139	7.19	215	4.65	291	3.44	367	2.72	443	2.26
140	7.14	216	4.63	292	3.42	368	2.72	444	2.25
141	7.09	217	4.61	293	3.41	369	2.71	445	2.25
142	7.04	218	4.59	294	3.40	370	2.70	446	2.24
143	6.99	219	4.57	295	3.39	371	2.70	447	2.24
144	6.94	220	4.55	296	3.38	372	2.69	448	2.23
145	6.90	221	4.52	297	3.37	373	2.68	449	2.23
146	6.85	222	4.50	298	3.36	374	2.67	450	2.22
147	6.80	223	4.48	299	3.34	375	2.67	451	2.22
148	6.76	224	4.46	300	3.33	376	2.66	452	2.21
149	6.71	225	4.44	301	3.32	377	2.65	453	2.21
150	6.67	226	4.42	302	3.31	378	2.65	454	2.20
151	6.62	227	4.41	303	3.30	379	2.64	455	2.20
152	6.58	228	4.39	304	3.29	380	2.63	456	2.19
153	6.54	229	4.37	305	3.28	381	2.62	457	2.19
154	6.49	230	4.35	306	3.27	382	2.62	458	2.18
155	6.45	231	4.33	307	3.26	383	2.61	459	2.18
156	6.41	232	4.31	308	3.25	384	2.60	460	2.17
157	6.37	233	4.29	309	3.24	385	2.60	461	2.17
158	6.33	234	4.27	310	3.22	386	2.59	462	2.16
159	6.29	235	4.26	311	3.22	387	2.58	463	2.16
160	6.25	236	4.24	312	3.21	388	2.58	464	2.16
161	6.21	237	4.22	313	3.19	389	2.57	465	2.15
162	6.17	238	4.20	314	3.18	390	2.56	466	2.15
163	6.13	239	4.18	315	3.17	391	2.56	467	2.14
164	6.10	240	4.17	316	3.16	392	2.55	468	2.14
165	6.06	241	4.15	317	3.15	393	2.54	469	2.13
166	6.02	242	4.13	318	3.14	394	2.54	470	2.13
167	5.99	243	4.12	319	3.13	695	2.53	471	2.12
168	5.95	244	4.10	320	3.12	396	2.52	472	2.12
169	5.92	245	4.08	321	3.11	397	2.52	473	2.11
170	5.88	246	4.07	322	3.10	398	2.51	474	2.11
171	5.84	247	4.05	323	3.10	399	2.51	475	2.11
172	5.81	248	4.03	324	3.09	400	2.50	476	2.10
173	5.78	249	4.02	325	3.08	401	2.49	477	2.10
174	5.75	250	4.00	326	3.07	402	2.49	478	2.09
175	5.71	251	3.98	327	3.06	403	2.48	479	2.09

TABLE FOR THE CONVERSION OF RATES INTO GRADIENTS—Continued.

Rate.	Gradient.	Rate.	Gradient.	Rate.	Gradient.	Rate.	Gradient.	Rate.	Gradient.
Feet.	°	Feet.	°	Feet.	°	Feet.	°	Feet.	°
480	2.08	556	1.80	632	1.58	708	1.41	784	1.28
481	2.08	557	1.80	633	1.58	709	1.41	785	1.27
482	2.07	558	1.79	634	1.58	710	1.41	786	1.27
483	2.07	559	1.79	635	1.57	711	1.41	787	1.27
484	2.07	560	1.79	636	1.57	712	1.40	788	1.27
485	2.06	561	1.78	637	1.57	713	1.40	789	1.27
486	2.06	562	1.78	638	1.57	714	1.40	790	1.27
487	2.05	563	1.78	639	1.56	715	1.40	791	1.26
488	2.05	564	1.77	640	1.56	716	1.40	792	1.26
489	2.04	565	1.77	641	1.56	717	1.39	793	1.26
490	2.04	566	1.77	642	1.56	718	1.39	794	1.26
491	2.04	567	1.76	643	1.56	719	1.39	795	1.26
492	2.03	568	1.76	644	1.55	720	1 39	796	1.26
493	2.03	569	1.76	645	1.55	721	1.39	797	1.25
494	2.02	570	1.75	646	1.55	722	1.39	798	1.25
495	2.02	571	1.75	647	1.55	723	1.38	799	1.25
496	2.02	572	1.75	648	1.54	724	1.38	800	1.25
497	2.01	573	1.75	649	1.54	725	1.38	801	1.25
498	2.01	574	1.74	650	1.54	726	1.38	802	1.25
499	2.00	575	1.74	651	1.54	727	1.38	803	1.25
500	2.00	576	1.74	652	1.53	728	1.37	804	1.24
501	2.00	577	1.73	653	1.53	729	1.37	805	1.24
502	1.99	578	1.73	654	1.53	730	1.37	806	1.24
503	1.99	579	1.73	655	1.53	731	1.37	807	1.24
504	1.98	580	1.72	656	1.52	732	1.37	808	1.24
505	1.98	581	1.72	657	1.52	733	1.36	809	1.24
506	1.98	582	1.72	658	1.52	734	1.36	810	1.23
507	1.97	583	1.72	659	1.52	735	1.36	811	1.23
508	1.97	584	1.71	660	1.52	736	1.36	812	1.23
509	1.96	585	1.71	661	1.51	737	1.36	813	1.23
510	1.96	586	1.71	662	1.51	738	1.36	814	1.23
511	1.96	587	1.70	663	1.51	739	1.35	815	1.23
512	1.95	588	1.70	664	1.51	740	1.35	816	1.23
513	1.95	589	1.70	665	1.50	741	1.35	817	1.22
514	1.95	590	1.69	666	1.50	742	1.35	818	1.22
515	1.94	591	1.69	667	1.50	743	1.35	819	1.22
516	1.94	592	1.69	668	1.50	744	1.34	820	1.22
517	1.93	593	1.69	669	1.49	745	1.34	821	1.22
518	1.93	594	1.68	670	1.49	746	1.34	822	1.22
519	1.93	595	1.68	671	1.49	747	1.34	823	1.22
520	1.92	596	1.68	672	1.49	748	1.34	824	1.21
521	1.92	597	1.68	673	1.49	749	1.34	825	1.21
522	1.92	598	1.67	674	1.48	750	1.33	826	1.21
523	1.91	599	1.67	675	1.48	751	1.33	827	1.21
524	1.91	600	1.67	676	1.48	752	1.33	828	1.21
525	1.90	601	1.66	677	1.48	753	1.33	829	1.21
526	1.90	602	1.66	678	1.47	754	1.33	830	1.20
527	1.90	603	1.66	679	1.47	755	1.32	831	1.20
528	1.89	604	1.66	680	1.47	756	1.32	832	1.20
529	1.89	605	1.65	681	1.47	757	1.32	833	1.20
530	1.89	606	1.65	682	1.47	758	1.32	834	1.20
531	1.88	607	1.65	683	1.46	759	1.32	835	1.20
532	1.88	608	1.64	684	1.46	760	1.32	836	1.20
533	1.88	609	1.64	685	1.46	761	1.31	837	1.19
534	1.87	610	1.64	686	1.46	762	1.31	838	1.19
535	1.87	611	1.64	687	1.46	763	1.31	839	1.19
536	1.87	612	1.63	688	1.45	764	1.31	840	1.19
537	1.86	613	1.63	689	1.45	765	1.31	841	1.19
538	1.86	614	1.63	690	1.45	766	1.31	842	1.19
539	1.86	615	1.63	691	1.45	767	1.30	843	1.19
540	1.85	616	1.62	692	1.45	768	1.30	844	1.18
541	1.85	617	1.62	693	1.44	769	1.30	845	1.18
542	1.85	618	1.62	694	1.44	770	1.30	846	1.18
543	1.84	619	1.62	695	1.44	771	1.30	847	1.18
544	1.84	620	1.61	696	1.44	772	1.30	848	1.18
545	1.83	621	1.61	697	1.43	773	1.29	849	1.18
546	1.83	622	1.61	698	1.43	774	1.29	850	1.18
547	1.83	623	1.61	699	1.43	775	1.29	851	1.18
548	1.82	624	1.60	700	1.43	776	1.29	852	1.17
549	1.82	625	1.60	701	1.43	777	1.29	853	1.17
550	1.82	626	1.60	702	1.42	778	1.29	854	1.17
551	1.81	627	1.59	703	1.42	779	1.28	855	1.17
552	1.81	628	1.59	704	1.42	780	1.28	856	1.17
553	1.81	629	1.59	705	1.42	781	1.28	857	1.17
554	1.81	630	1.59	706	1.42	782	1.28	858	1.17
555	1.80	631	1.58	707	1.41	783	1.28	859	1.16

TABLE FOR THE CONVERSION OF RATES INTO GRADIENTS—Continued.

Rate.	Gradient.	Rate.	Gradient.	Rate.	Gradient.	Rate.	Gradient.	Rate.	Gradient.
Feet.	°	Feet.	°	Feet.	°	Feet.	°	Feet.	°
860	1.16	889	1.12	917	1.09	945	1.06	973	1.03
861	1.16	890	1.12	918	1.09	946	1.06	974	1.03
862	1.16	891	1.12	919	1.09	947	1.06	975	1.03
863	1.16	892	1.12	920	1.09	948	1.05	976	1.02
864	1.16	893	1.12	921	1.09	949	1.05	977	1.02
865	1.16	894	1.12	922	1.08	950	1.05	978	1.02
866	1.15	895	1.12	923	1.08	951	1.05	979	1.02
867	1.15	896	1.12	924	1.08	952	1.05	980	1.02
868	1.15	897	1.11	925	1.08	953	1.05	981	1.02
869	1.15	898	1.11	926	1.08	954	1.05	982	1.02
870	1.15	899	1.11	927	1.08	955	1.05	983	1.02
871	1.15	900	1.11	928	1.08	956	1.05	984	1.02
872	1.15	901	1.11	929	1.08	957	1.04	985	1.02
873	1.15	902	1.11	930	1.08	958	1.04	986	1.01
874	1.14	903	1.11	931	1.07	959	1.04	987	1.01
875	1.14	904	1.11	932	1.07	960	1.04	988	1.01
876	1.14	905	1.10	933	1.07	961	1.04	989	1.01
877	1.14	906	1.10	934	1.07	962	1.04	990	1.01
878	1.14	907	1.10	935	1.07	963	1.04	991	1.01
879	1.14	908	1.10	936	1.07	964	1.04	992	1.01
880	1.14	909	1.10	937	1.07	965	1.04	993	1.01
881	1.14	910	1.10	938	1.07	966	1.04	994	1.01
882	1.13	911	1.10	939	1.06	967	1.03	995	1.01
883	1.13	912	1.10	940	1.06	968	1.03	996	1.00
884	1.13	913	1.10	941	1.06	969	1.03	997	1.00
885	1.13	914	1.09	942	1.06	970	1.03	998	1.00
886	1.13	915	1.09	943	1.06	971	1.03	999	1.00
887	1.13	916	1.09	944	1.06	972	1.03	1000	1.00
888	1.13								

SUMMARY OF OBSERVATIONS.

TEMPERATURE.

The mean rate of diminution of temperature with increase of altitude, as determined from 1,217 ascensions and 3,838 observations, taken at elevations of 1,000 feet or more, was 5.0° for each 1,000 feet, or 0.4° less than the true adiabatic rate. The largest gradient, 7.4° per thousand feet, was found up to 1,000 feet, and thereafter there was a steady decrease up to 5,000 feet, the rate of decrease becoming less as the altitude increased. The gradient up to 5,000 feet was 3.8° per thousand feet. Above this altitude there is a tendency toward a slow rise, but the lack of a sufficient number of observations above 6,000 feet forbids a definite statement to that effect. The morning gradients were also greatest up to 1,000 feet, and least up to 5,000 feet, and the rate of decrease was about the same as the mean rate, the curves showing a very close agreement in this respect. (See plates 2 and 4.) The average morning gradient was 4.8° per thousand feet. The afternoon gradients were larger, but not decidedly so, the average value being 5.8° per thousand feet. The greatest rate of decrease is still found at 1,000 feet, and the least up to 5,000 feet, if the few observations at 7,000 feet are not considered as of equal weight. The morning, afternoon, and mean gradients for the different elevations from 1,000 to 8,000 feet, inclusive, are given in the following table:

DECREASE OF TEMPERATURE FOR EACH RESPECTIVE 1,000 FEET OF ALTITUDE.

	1,000 feet.	1,500 feet.	2,000 feet.	3,000 feet.	4,000 feet.	5,000 feet.	6,000 feet.	7,000 feet.	8,000 feet.	Mean.
	°	°	°	°	°	°	°	°	°	°
Morning	7.2	5.5	4.8	4.0	3.7	3.7	3.9	3.4	3.0	4.8
Afternoon	7.5	6.4	6.0	5.5	4.9	4.3	4.5	3.5	4.9	5.8
Mean	7.4	5.8	5.2	4.4	4.0	3.8	4.1	3.4	4.0	5.0

When the stations of observation were grouped according to their geographical locations it was found that the mean rate of temperature decrease with increase of altitude was much greater in the central Mississippi watershed than in the Upper Lake region, the central West, or the extreme East as represented by the single station at Washington. It was 5.8° per thousand feet, as compared with 4.7° for the central West, 4.6° for the Upper Lake region, and 3.6° for the Atlantic coast. It will be observed also that there is a very close agreement between the means for the Upper Lake region, those for the central West, and the grand mean, and a marked deficiency on the Atlantic coast, where the gradient was 1.4° per thousand feet less than the mean rate. These general statements apply also to the morning results. In the afternoon, however, the differences were quite small, the extreme difference being 0.6° per thousand feet. The maximum gradient, 6.1°, was found in the central Mississippi Basin, and the minimum, 5.4°, in the Upper Lake region. The morning, afternoon, and mean results for the various districts are shown in the following table:

GRADIENT PER THOUSAND FEET.

District.	Morning.	After-noon.	Mean.
	°	°	°
Atlantic coast	3.4	6.0	3.6
Central Mississippi watershed	5.8	6.1	5.8
Upper Lake region	4.5	5.4	4.6
Central West	4.3	5.6	4.7

RELATIVE HUMIDITY.

The relative humidities at and above the surface of the earth differed but little except at 7,000 feet, where the surface humidity was 11 per cent less than that above. With this exception the greatest difference was 3 per cent, and, except at 2,000 and 8,000 feet, the upper air percentages were the lower. The mean result obtained from all the observations showed 60 per cent at the surface and 58 per cent above, a difference of 2 per cent.

The stations at which there were marked differences were Washington, where the mean difference was 14 per cent; Omaha, where it was 29 per cent; Springfield, Ill., where it was 21 per cent, and Fort Smith, where it was 12 per cent, the surface humidity being the higher except at Fort Smith. At the remaining thirteen stations except Lansing, the upper air humidity equaled or exceeded that at the surface, but the difference at no place exceeded 10 per cent. At nine stations the difference was 5 per cent or less. (See table of mean relative humidity, p. 20.)

VAPOR PRESSURE.

The vapor pressures are expressed in percentages obtained by the formula $\frac{p}{p^\circ}$, in which "p" represents the vapor pressure at any given altitude, and "p°" that observed simultaneously at the earth's surface. The mean of the percentages thus obtained was 59, and there was a steady, though not by any means uniform, decrease with increase of altitude. The percentage at 1,500 feet was 82, and at 8,000 feet, 44. The decrease was most rapid between 2,000 and 5,000 feet, where it averaged 9 per cent for each 1,000 feet. The decrease between 5,000 and 6,000 feet was only 3 per cent, while between 6,000 and 7,000 feet it was 10 per cent. The lowest percentage, 52, was found at Omaha, and the highest, 77, at Pierre. (See table on vapor pressure, p. 20.

A comparative statement of the results obtained from the kite, balloon, and mountain observations is given herewith. In obtaining these results the records of 1,123 kite ascensions were used. There were 4 balloon ascensions by Hammon and 2 by Hazen. It is not known how many were made by Hann, nor how many mountain observations were taken by him.

DIMINUTION OF VAPOR PRESSURE WITH ALTITUDE.

VALUE OF $\frac{p}{p^\circ}$ FOR EACH RESPECTIVE 1,000 FEET OF ALTITUDE.

Character of observations.	1,500 feet.	2,000 feet.	3,000 feet.	4,000 feet.	5,000 feet.	6,000 feet.	7,000 feet.	8,000 feet.	Mean.
Kite	0.82	0.78	0.70	0.61	0.52	0.49	0.39	0.44	0.59
Balloon (Hammon)	0.96	0.96	0.87	0.68	0.44	0.59			0.75
Balloon (Hazen)	0.89	0.83	0.80	0.78	0.67	0.46	0.44		0.70
Balloon (Hann)	0.84	0.80	0.66	0.61	0.60	0.54	0.41	0.37	0.60
Mountain (Hann)	0.83	0.81	0.80	0.66	0.61	0.58	0.55	0.47	0.66

TEMPERATURE GRADIENTS.

MORNING.

[Bold-faced figures represent degrees per thousand feet; light-faced figures represent number of observations.]

Station.	1,000 feet.	1,500 feet.	2,000 feet.	3,000 feet.	4,000 feet.	5,000 feet.	6,000 feet.	7,000 feet.	8,000 feet.	Mean.
Washington, D.C	4.6	3.9	3.6	3.2	3.0	2.8	3.1	3.0	3.0	3.4
	37	41	44	25	17	8	7	3	1	183
Cairo, Ill	10.5	6.4	5.7	4.3	4.4	4.1	......	......	......	5.9
	4	19	17	14	12	3				69
Cincinnati, Ohio......	11.6	6.3	5.5	5.7	5.6	4.7	3.7	......	......	6.2
	1	6	14	13	12	7	2			55
Fort Smith, Ark......	7.8	6.9	6.7	5.6	3.8	......	......	......	......	6.2
	9	9	9	3	2					32
Knoxville, Tenn	9.1	8.2	8.2	4.5	4.5	......	......	......	......	6.9
	1	2	2	4	1					10
Memphis, Tenn.......	8.7	6.0	4.6	3.3	3.6	3.6	......	......	......	5.0
	7	7	13	19	18	4				68
Springfield, Ill	6.8	5.2	4.5	4.6	3.8	3.7	......	......	......	4.8
	7	17	20	11	2	1				58
Cleveland, Ohio	5.9	4.1	3.5	3.5	4.1	4.1	4.3	......	......	4.2
	22	58	53	48	21	11	1			214
Duluth, Minn	4.7	4.3	3.8	4.1	4.3	4.2	4.8	......	......	4.3
	39	47	51	32	20	8	3			200
Lansing, Mich........	7.8	6.0	4.8	3.9	3.8	3.8	......	......	......	5.0
	19	19	20	29	20	10				117
Sault Ste. Marie, Mich.	6.0	5.4	4.8	3.9	3.3	3.0	......	......	......	4.4
	6	22	28	27	20	13				116
Dodge City, Kans	6.4	5.0	4.5	3.3	2.6	2.9	3.1	3.0	......	3.8
	76	83	79	79	49	19	6	1		392
Dubuque, Iowa.......	7.4	5.4	4.4	3.2	3.5	3.4	......	......	......	4.6
	9	17	29	24	9	5				93
North Platte, Nebr....	6.2	5.7	4.8	4.2	3.7	4.2	......	......	......	4.8
	34	63	52	38	22	7				216
Omaha, Nebr	....	4.6	4.2	3.1	2.6	2.9	3.4	4.1	....	3.6
		19	34	37	26	14	5	2		137
Pierre, S. Dak	5.3	4.8	4.3	3.9	3.5	4.2	4.0	......	......	4.3
	71	83	71	53	34	17	2			331
Topeka, Kans	7.0	5.4	4.1	3.3	3.3	3.8	4.5	......	......	4.5
	34	32	43	48	17	9	1			184
Mean	7.2	5.5	4.8	4.0	3.7	3.7	3.9	3.4	3.0	4.8
	376	544	579	504	302	136	27	6	1	2,475

AFTERNOON.

Washington, D.C	8.1	6.5	5.8	5.5	4.9	5.1	......	......	......	6.0
	14	11	9	3	2	1				40
Cairo, Ill	8.0	7.0	6.5	5.8	5.2	4.7	......	......	......	6.2
	2	14	14	9	7	1				47
Cincinnati, Ohio......	13.8	6.6	8.8	6.0	5.5	5.0	5.2	......	......	7.3
	2	1	10	6	4	2	1			26
Fort Smith, Ark	5.2	7.1	6.7	6.0	......	......	......	......	......	6.2
	3	9	9	5						26
Knoxville, Tenn	8.2	5.8	6.0	6.3	5.5	......	......	......	......	6.3
	7	6	6	4	1					24
Memphis, Tenn	7.2	7.2	5.4	5.2	5.0	3.4	......	......	......	5.6
	9	15	14	7	3	1				49
Springfield, Ill	8.3	5.9	5.5	4.2	4.1	3.7	3.6	......	......	5.0
	8	35	33	17	13	8	2			116
Cleveland, Ohio	4.9	5.0	6.0	......	......	......	......	......	......	5.3
	7	4	2							13
Duluth, Minn.........	8.0	5.9	5.8	5.1	4.2	3.3	4.2	......	......	4.9
	22	24	32	28	12	5	2			125
Lansing, Mich	5.8	6.2	3.9	5.3	5.3	......	......	......	......	5.3
	4	4	2	5	2					17
Sault Ste. Marie, Mich.	10.0	7.0	5.7	5.8	5.0	3.4	......	......	......	6.2
	1	20	18	12	11	2				64
Dodge City, Kans.....	6.2	5.6	5.4	4.8	4.4	3.7	3.4	3.5	4.9	4.7
	43	40	39	27	17	9	4	1	1	181
Dubuque, Iowa.......	6.4	6.5	4.9	4.2	2.8	2.8	......	......	......	4.6
	10	15	11	11	6	2				55
North Platte, Nebr....	7.3	7.0	6.6	5.7	5.0	5.3	5.4	......	......	6.0
	36	99	81	64	22	6	1			309
Omaha, Nebr	....	7.1	6.4	6.0	5.4	5.2	5.3	......	......	5.9
		10	16	8	8	5	1			48
Pierre, S. Dak	7.5	6.4	6.2	6.1	5.7	6.0	......	......	......	6.3
	25	22	20	12	4	2				85
Topeka, Kans	7.8	6.9	6.1	5.5	4.9	4.7	......	......	......	6.0
	34	33	33	24	9	2				135
Mean	7.5	6.4	6.0	5.5	4.9	4.3	4.5	3.5	4.9	5.8
	227	362	349	242	121	46	11	1	1	1,360

TEMPERATURE GRADIENT   Continued.

MEAN.

Station.	1,000 feet.	1,500 feet.	2,000 feet.	3,000 feet.	4,000 feet.	5,000 feet.	6,000 feet.	7,000 feet.	8,000 feet.	Mean.
Washington, D.C.	5.6	4.4	4.0	3.5	3.2	3.0	3.1	3.0	3.0	3.6
	51	52	53	28	19	9	7	3	1	223
Cairo, Ill	9.7	6.6	6.0	4.9	4.7	4.3	....	....	....	6.0
	6	33	31	23	19	4				116
Cincinnati, Ohio	13.0	6.3	6.9	5.8	5.6	4.7	4.2	....	....	6.6
	3	7	24	19	16	9	3			81
Fort Smith, Ark	7.2	7.0	6.7	5.8	3.8					6.2
	12	18	18	8	2					58
Knoxville, Tenn	8.4	6.2	6.6	5.4	5.0					6.3
	8	8	8	8	2					34
Memphis, Tenn	7.8	6.8	5.0	3.8	3.7	3.5				5.1
	16	22	27	26	21	5				117
Springfield, Ill	7.6	5.7	5.1	4.4	4.0	3.7	3.6			4.9
	15	52	53	28	15	9	2			174
Cleveland, Ohio	5.7	4.1	3.6	3.5	4.1	4.1	4.3			4.2
	29	62	55	48	21	11	1			227
Duluth, Minn.........	5.2	4.8	4.6	4.6	4.3	3.8	4.6			4.6
	61	71	83	60	32	13	5			325
Lansing, Mich	7.5	6.0	4.7	4.1	3.9	3.8				5.0
	23	23	22	34	22	10				134
Sault Ste. Marie, Mich.	6.6	6.2	5.2	4.5	3.9	3.0				4.9
	7	42	46	39	31	15				180
Dodge City, Kans.....	6.3	5.2	4.8	3.7	3.1	3.2	3.2	3.2	4.9	4.2
	119	123	118	106	66	28	10	2	1	573
Dubuque, Iowa.......	6.9	5.9	4.6	3.5	3.2	3.3				4.6
	19	32	40	35	15	7				148
North Platte, Nebr....	6.8	6.5	5.9	5.2	4.4	4.7	5.4			5.6
	70	162	133	102	44	13	1			525
Omaha, Nebr	....	5.4	4.9	3.6	3.2	3.5	3.8	4.1		4.1
		29	50	45	34	19	6	2		185
Pierre, S. Dak	5.9	5.1	4.8	4.3	3.7	4.4	4.0			4.6
	96	105	91	65	38	19	2			416
Topeka, Kans.........	7.4	6.2	4.9	4.0	3.8	3.9	4.5			5.0
	68	65	76	72	26	11	1			319
Mean ....{	7.4	5.8	5.2	4.4	4.0	3.8	4.1	3.4	4.0	5.0
	603	906	928	746	423	182	38	7	2	3,835

TEMPERATURE GRADIENTS BY GEOGRAPHICAL DISTRICTS.

MORNING.

District.	1,000 feet.	1,500 feet.	2,000 feet.	3,000 feet.	4,000 feet.	5,000 feet.	6,000 feet.	7,000 feet.	8,000 feet.	Mean.
Atlantic coast	4.6	3.9	3.6	3.2	3.0	2.8	3.1	3.0	3.0	3.4
Central Mississippi watershed..........	9.1	6.5	5.9	4.7	4.3	4.0	3.7	....	....	5.8
Upper Lake Region ...	6.1	5.0	4.2	3.8	3.9	3.8	4.6	4.2	....	4.5
Central West..........	6.5	5.2	4.4	3.5	3.2	3.6	3.8	3.6		4.3

AFTERNOON.

District.	1,000 feet.	1,500 feet.	2,000 feet.	3,000 feet.	4,000 feet.	5,000 feet.	6,000 feet.	7,000 feet.	8,000 feet.	Mean.
Atlantic coast	8.1	6.5	5.8	5.5	4.9	5.1	....	....	....	6.0
Central Mississippi watershed..........	8.4	6.5	6.5	5.6	5.1	4.2	4.4	....	....	6.1
Upper Lake Region ...	6.7	6.0	5.4	5.4	4.8	3.4	4.2	3.8	....	5.4
Central West..........	7.0	6.6	5.9	5.3	4.7	4.6	3.7	3.5	4.9	5.6

MEAN.

District.	1,000 feet.	1,500 feet.	2,000 feet.	3,000 feet.	4,000 feet.	5,000 feet.	6,000 feet.	7,000 feet.	8,000 feet.	Mean.
Atlantic coast	5.6	4.4	4.0	3.5	3.2	3.0	3.1	3.0	3.0	3.6
Central Mississippi watershed..........	9.0	6.4	6.0	5.0	4.5	4.0	3.9	....	....	5.8
Upper Lake Region ...	6.2	5.3	4.5	4.2	4.1	3.7	4.4	4.0		4.6
Central West..........	6.6	5.7	5.0	4.0	3.6	3.8	4.2	3.6	4.9	4.7

MEAN RELATIVE HUMIDITY.

[S = at ground; A = above.]

Station.	1,500 feet.		2,000 feet.		3,000 feet.		4,000 feet.		5,000 feet.		6,000 feet.		7,000 feet.		8,000 feet.		Mean.	
	S.	A.	S.	A.	S.	A.	S.	A.	S.	A.	S.	A.	S.	A.	S.	A.	S.	A.
Washington, D. C	76	68	75	67	77	64	75	61	80	60	78	66	73	59	79	58	77	63
Cairo, Ill	68	67	65	66	63	64	62	62	60	65	....	....	....	....	....	....	64	65
Cincinnati, Ohio..........	70	67	65	68	64	63	66	68	65	64	63	70	....	....	....	....	66	67
Fort Smith, Ark	63	72	63	76	63	82	71	79	....	....	....	....	....	....	....	....	65	77
Knoxville, Tenn	55	63	61	71	72	75	64	64	....	....	....	....	....	....	....	....	63	·68
Memphis, Tenn	59	70	67	77	72	77	69	69	60	62	....	....	....	....	....	....	65	71
Springfield, Ill	61	45	60	42	57	36	46	24	53	26	50	24	....	....	....	....	54	33
Cleveland, Ohio	74	71	74	72	73	72	69	65	69	75	78	94	....	....	....	....	73	75
Duluth, Minn.............	70	72	66	69	60	65	62	64	61	57	58	57	54	70	....	....	62	65
Lansing, Mich	78	69	76	68	72	69	61	68	65	64	....	....	....	....	....	....	70	68
Sault Ste. Marie, Mich	80	71	73	69	73	69	63	68	59	73	....	....	....	....	....	....	70	70
Dodge City, Kans	57	52	57	58	5?	56	52	42	48	51	45	50	32	43	22	47	46	50
Dubuque, Iowa............	70	73	72	73	65	66	66	66	58	71	....	....	....	....	....	....	68	68
North Platte, Nebr	53	56	51	56	47	53	46	52	42	49	30	60	....	....	....	....	45	54
Omaha, Nebr	48	40	50	40	49	33	53	30	62	24	67	15	66	10	....	....	56	27
Pierre, S. Dak	57	63	56	61	55	66	52	60	43	58	48	64	....	....	....	....	52	62
Topeka, Kans	61	69	62	67	63	64	60	59	62	61	63	63	....	....	....	....	62	64
Mean	65	64	64	65	64	63	61	58	60	57	58	56	56	45	50	52	60	58

VAPOR PRESSURE $\left(\dfrac{P}{p°}\right)$

Station.	1,500 feet.	2,000 feet.	3,000 feet.	4,000 feet.	5,000 feet.	6,000 feet.	7,000 feet.	8,000 feet.	Mean.
Washington, D. C	0.87	0.82	0.66	0.60	0.54	0.46	0.45	0.34	0.59
Cairo, Ill	.71	.69	.63	.54	.54	....	....	....	.62
Cincinnati, Ohio..............	.73	.67	.57	.51	.49	.45	....	....	.57
Fort Smith, Ark	.82	.79	.74	.65	....	....	....	....	.75
Knoxville, Tenn	.83	.73	.66	.76	....	....	....	....	.74
Memphis, Tenn	.88	.86	.76	.61	.54	....	....	....	.73
Springfield, Ill	.74	.70	.63	.52	.48	.49	....	....	.59
Cleveland, Ohio	.83	.77	.68	.55	.55	.48	....	....	.64
Duluth, Minn.................	.82	.79	.74	.64	.57	.56	.45	....	.65
Lansing, Mich	.87	.83	.73	.56	.51	....	....	....	.70
Sault Ste. Marie, Mich	.83	.76	.71	.71	.44	....	....	....	.69
Dodge City, Kans	.85	.84	.80	.71	.61	.56	.51	.55	.68
Dubuque, Iowa................	.83	.79	.76	.56	.56	....	....	....	.72
North Platte, Nebr............	.78	.72	.66	.57	.40	.65	....	....	.63
Omaha, Nebr	.83	.80	.68	.56	.39	.23	.15	....	.52
Pierre, S. Dak	.90	.86	.75	.72	.69	.69	....	....	.77
Topeka, Kans.................	.85	.80	.68	.56	.52	.36	....	....	.63
Mean..................	.82	.78	.70	.61	.52	.49	.39	.44	.59

WASHINGTON, D. C. (ARLINGTON, VA.).

There were in all 87 ascensions and 223 observations, taken during the daytime only, at altitudes of 1,000 feet or over, and the highest altitude attained was 8,211 feet.

The general mean decrease in temperature was 3.6° per thousand feet. The average decrease from the ground up to 1,000 feet of altitude above the kite station was 5.6°; for the other altitudes it was as follows: 1,500 feet, 4.4° per thousand feet; 2,000 feet, 4.0°; 3,000 feet, 3.5°; 4,000 feet, 3.2°; 5,000 feet, 3.0°; 6,000 feet, 3.1°; 7,000 feet, 3.0°; 8,000 feet, 3.0°.

These results are graphically shown on Plate 3. They are the general mean values obtained from observations taken at all hours between 6 a. m. and 7 p. m., Eastern time. No distinction is now made between clear and cloudy days.

It will be noticed that there is a decrease in the gradient up to 4,000 feet, above which altitude the changes were very slight, but nevertheless with a decreasing tendency. The results obtained at altitudes above 4,000 feet are remarkably uniform, considering the paucity of observations and the wide divergence in the weather conditions. But 20 observations were made at altitudes of 5,000 feet or more, and of these 10 were made on clear days, 8 on days on which rain fell, and the remaining 2 on cloudy days without rain.

The negative gradients of temperature or "inversions" during the morning hours of course bear a direct relation to the amount of cloudiness and the velocity of the wind. Thirty-five cases were found with a fairly even distribution through the different months. The inversions, as a rule, extended to an altitude of 1,200 feet, and in over one-half the cases reached 1,500 feet. On May 12, after a clear night with a high relative humidity, there was an inversion of 4.5° up to 2,500 feet, and it did not cease until the kite reached an altitude of 4,100 feet. The most marked case occurred on June 21. The night and early morning up to 6 o'clock had been clear (becoming cloudy, however, by 7 o'clock), and the sun rose at 4.35 a. m. At 5.10 a. m., or thirty-five minutes after sunrise, the temperature at 866 feet elevation was 14° higher than at the ground, and was 10° higher at 5.20 a. m. at an altitude of 1,700 feet. The surface wind was from the west and its velocity 3 miles per hour. A very similar case occurred on June 25, when at 5.40 a. m., one hour and three minutes after sunrise, the temperature at 1,966 feet was 9° higher than at the surface. At this time the surface wind was from the southwest and its velocity also 3 miles per hour.

It is worthy of note that where the inversions were smallest, and incidentally the wind velocities greatest, the wind was always from some northerly point, usually the northwest; and, vice versa, with strong inversions and light winds, the latter were almost always from some southerly point, usually the southwest.

It is interesting also to note the gradients at different times of the day. In the morning the mean rate was 3.4° per thousand feet, the greatest being 4.6° at 1,000 feet, and the least 2.8° at 5,000 feet. The curve in this case fairly approximates that for the mean result, and is given on Plate 1.

In the afternoon, up to a limit of 5,000 feet in altitude, the rate of decrease so nearly approaches the adiabatic rate of 1° for 185 feet, or 5.4° per thousand feet, that the two can be considered as practically one, the mean value obtained being 6.0° per thousand feet, or 0.6° greater than the true adiabatic rate. The lowest rate, 4.9° per thousand feet, was noted at 4,000 feet, and the highest, 8.1° per thousand feet, at 1,000 feet. The corresponding curve for these figures is given on Plate 2.

At all times where clouds were present there was a great decrease in the gradient, resulting in some cases in a complete arrest of the temperature fall, and in others a marked rise, notwithstanding the increased altitude. On July 13, while the kite was in the clouds at altitudes between 2,020 and 2,456 feet, a rise of 5.0° occurred. The extreme range of temperature while the kite was in the clouds was 1.7° for 1,300 feet difference in altitude, or at the rate of 1.3° per thousand feet.

This rise in temperature in the clouds was also noted in several other cases. On June 5, while no rise in temperature occurred, the fall was almost entirely checked, the total decrease for 1,743 feet of ascent having been but 1.1°, or at the rate of 0.6° per thousand feet. An examination of the temperature record for October 8 disclosed some peculiar cloud effects. The flight took place in the morning, the sky being practically obscured by stratus clouds. In this case the rise in temperature began at an altitude of 2,300 feet, about 1,300 feet below the clouds, and continued up to 3,100 feet, after which there was a fall of 2.5° up to 3,600 feet, when the kite entered the clouds, and no change while it remained there. The gradient during the entire flight was very small, occasionally amounting to a complete inversion, particularly at the height of 3,100 feet, when the inversion was as much as 0.6° per thousand feet.

The wind velocity appears to have no effect upon the temperature gradient, inversions not considered, except that during cloudy weather there was usually a marked decrease in the gradient when the surface velocity was about 5 miles per hour or less, and that above correspondingly small as evidenced by the pull on the kite wire.

No deductions regarding the diurnal range of temperature in the upper air were made, owing to the very limited time during which the kite remained at a fixed elevation.

WIND.

The differences in wind direction were indicated by the changes in the azimuths of the kite. These show that in a great majority of cases the general directions above and at the surface are practically the same, the differences being confined to a tendency in the kite to a deflection toward the right. This deflection frequently increased with the altitude, but rarely equaled 90 degrees. In a few instances, not over five or six, the kite was deflected toward the left, but not to any great extent except in one case, on May 27, when the deflection toward the left was about 140 degrees. It should also be stated that in a majority of the cases where the kite was deflected toward the left the velocity of the wind diminished with increase of altitude, as shown by the pull on the kite wire.

RELATIVE HUMIDITY AND VAPOR PRESSURE.

The relative humidity records show that in a great majority of instances, under normal clear conditions, the change with altitude is largely a question of wind direction, or really of the relative positions of the high and low pressure areas. With northerly winds, particularly northwesterly, there was as a rule a marked decrease in the relative humidity with increase of altitude, while with southerly winds the rule was no change or a steady increase, usually the latter, and with easterly winds a decided increase. The drying character of the northwest winds was sometimes noticed at different intervals during the same ascension as the upper currents changed direction from time to time.

In the presence of clouds and rain, or with the latter a few hours before or after the flight, an increase was the rule, but frequently not until after an altitude of 2,500 or 3,000 feet had been reached. But one case was recorded of total saturation while the kite was in the clouds. In one other case, on October 8, the relative humidity fell 22 per cent in fifteen minutes just previous to entering the clouds, remaining at the low point until the kite left the clouds, when it again rose to its former height.

The vapor pressure results for Washington are given below, and are expressed in fractions of the vapor pressure observed simultaneously at the earth's surface $\left(\dfrac{p}{p^o}\right)$ in which "P" represents the vapor pressure at any given altitude and "p^o" that at the earth's surface.

DIMINUTION OF VAPOR PRESSURE WITH ALTITUDE.

VALUE OF $\dfrac{p}{p^o}$ AT EACH RESPECTIVE 1,000 FEET OF ALTITUDE.

1,500 feet.	2,000 feet.	3,000 feet.	4,000 feet.	5,000 feet.	6,000 feet.	7,000 feet.	8,000 feet.
0.87	0.82	0.66	0.60	0.54	0.46	0.45	0.34

The computations of the gradients with the general summary of the results are given in the following table.

The first column under each altitude shows the rate of change of altitude for 1° F. change in temperature, while the second column shows the gradient of the temperature for each 1,000 feet of altitude.

TEMPERATURE.

CLEAR.

Date	1,000 feet		1,500 feet		2,000 feet		3,000 feet		4,000 feet		5,000 feet		6,000 feet		7,000 feet		8,000 feet		Mean	
	Rate	Gradient	Rate	Gradient	Rate	Gradient	Rate	Gradient	Rate	Gradient	Rate	Gradient	Rate	Gradient	Rate	Gradient	Rate	Gradient	Rate	Gradient
1898.																				
May 9, a. m...			848	1.2	582	1.7														
10, p. m...			165	6.1																
12, a. m...			350	2.9	192	5.2					272	3.7	539	1.9						
12, a. m...													296	3.4	446	2.2	335	3.0		
13, a. m...					142	7.0														
14, a. m...			350	2.9	192	5.2														
18, p. m...	105	9.5	163	6.1																
19, a. m...	500	2.0	339	2.9					510	2.0										
24, a. m...	467	2.1	1,180	0.9					223	4.5										
30, p. m...	669	1.2	244	4.1	230	4.3														
31, a. m...					1,140	0.9	569	1.8												
31, a. m...							277	3.6												
June 3, p. m...	173	5.8	190	5.3	190	5.3	169	5.9												
4, a. m...	170	5.9	336	3.0	262	3.8														
8, a. m...	120	8.3																		
8, p. m...	126	7.9	118	8.5	139	7.2														
9, a. m...			83	13.0	111	9.0			180	5.6	218	4.6								
12, a. m...	212	4.7																		
14, a. m...							754	1.3	553	1.8	492	2.0	339	2.9						
15, a. m...									250	4.0										
21, a. m...											272	3.7	299	3.3						
22, a. m...					400	2.5			271	3.7	214	4.7	272	3.7						
23, a. m...					247	4.0					296	3.4	291	3.4						
24, a. m...					570	1.8														
25, a. m...											340	2.9	312	3.2	297	3.4				
26, a. m...											657	1.5	377	2.7						
29, a. m...	141	7.1	170	5.9	158	6.5														
July 2, p. m...									219	4.6										
8, a. m...	122	8.2																		
9, a. m...	394	2.5	159	6.3	200	5.0			173	5.8	348	2.9	354	2.8	290	3.4	283	3.5		
10, a. m...									398	2.5										
11, a. m...							234	4.3												
15, p. m...	91	11.0																		
20, p. m...	75	13.3																		
21, a. m...	834	1.2	629	1.6	638	1.6														
28, a. m...	278	3.7	285	3.5	282	3.5														
30, a. m...	250	4.0	264	3.8																
Aug. 2, a. m...	299	3.3																		
3, a. m...	104	9.6	130	7.7																
4, a. m...	246	4.1	404	2.5			784	1.4	558	1.8	305	3.3								
5, a. m...	1,000	1.0	703	1.4			647	1.5	333	3.0										
7, p. m...	141	7.1	166	6.0																
16, p. m...	119	8.4	142	7.0	163	6.1														
17, a. m...			122	8.2	135	7.4														
17, p. m...	84	11.9	98	10.2	152	6.6														
17, p. m...																				
18, a. m...					146	6.8														
18, p. m...	137	7.3																		
22, p. m...	111	9.0																		
22, a. m...	724	1.4																		
24, p. m...	105	9.5	136	7.4	144	6.9	167	6.0	188	5.3	195	5.1								
27, a. m...	438	2.3	403	2.5	211	4.7														
29, a. m...	88	11.4	418	2.4	135	7.4														
30, a. m...	139	7.2																		
Sept. 2, a. m...							844	1.2												
26, p. m...	123	8.1	180	5.6																
27, a. m...			630	1.6			276	3.6	254	3.9										
29, a. m...	116	8.6																		
Oct. 9, a. m...	236	4.2	310	3.2			260	3.8												
10, p. m...			125	8.0			168	6.0												
12, a. m...							1,161	0.9	283	3.5										
15, a. m...					483	2.1	510	2.1	408	2.5	288	3.5								
17, a. m...	749	1.3			290	3.4														
19, a. m...	838	1.2					695	1.4												
22, a. m...	250	4.0			501	2.0	792	1.3												
23, a. m...					646	1.5	277	3.6												
24, a. m...					119	8.4	151	6.6												
25, a. m...											1,956	0.5								
25, a. m...											848	1.2	1,131	0.9						
27, a. m...	357	2.8	423	2.4			219	4.6	258	3.9										
31, a. m...			2,020	0.5			529	1.9												

CLOUDY.

Date	1,000 feet		1,500 feet		2,000 feet		3,000 feet		4,000 feet		5,000 feet		6,000 feet		7,000 feet		8,000 feet		Mean	
1898.																				
May 6, a. m...	617	1.6	924	1.1																
11, a. m...							2,370	0.4												
16, a. m...									250	4.0			298	3.4	291	3.4				
25, a. m...					546	1.4														
26, a. m...	254	3.9																		
27, a. m...	122	8.2	165	6.1	179	5.6			208	4.8										

Temperature—Cloudy—Continued.

Date.	1,000 feet. Rate.	Gradient.	1,500 feet. Rate.	Gradient.	2,000 feet. Rate.	Gradient.	3,000 feet. Rate.	Gradient.	4,000 feet. Rate.	Gradient.	5,000 feet. Rate.	Gradient.	6,000 feet. Rate.	Gradient.	7,000 feet. Rate.	Gradient.	8,000 feet. Rate.	Gradient.	Mean. Rate.	Gradient.
1898.																				
June 5, a.m....	314	3.2	234	4.3	239	4.2	271	3.7												
5, a.m....			184	6.1																
11, a.m....	187	5.3	109	5.0																
16, a.m....			252	4.0	266	3.8	313	3.2	262	3.8										
July 4, a.m....									1,771	0.6	326	3.1								
5, a.m....	380	2.6	251	4.0	249	4.0														
13, a.m....	346	2.9					594	1.7												
13, a.m....			255	3.9	594	1.7	442	2.3												
15, a.m....			496	2.0	358	2.8														
22, a.m....			1,650	0.6	294	3.4														
23, a.m....	310	3.2	225	4.4																
25, a.m....	474	2.1	141	6.2																
Aug. 9, a.m....	114	8.8																		
10, a.m....					461	2.3	600	1.7	529	1.9			301	3.3						
10, a.m....					207	4.8														
14, a.m....							265	3.8												
19, a.m....	147	6.8																		
19, a.m....	175	5.7																		
27, a.m....					93	10.8														
Oct. 2, a.m....	599	1.7																		
6, a.m....					1,371	0.7	221	4.5												
8, a.m....					1,046	1.0					1,064	0.9								
18, a.m....	348	2.9	266	3.8	262	3.8														

SUMMARY.

CLEAR.

	1,000	1,500	2,000	3,000	4,000	5,000	6,000	7,000	8,000	Mean
Morning......	4.8	3.8	3.7	3.4	3.1	2.8	3.0	2.8	3.0	3.4
Afternoon....	8.1	6.5	5.8	5.5	4.9	5.1				6.0
Mean...	6.0	4.6	4.2	3.7	3.4	3.0	3.0	2.8	3.0	3.7

CLOUDY.

	1,000	1,500	2,000	3,000	4,000	5,000	6,000	7,000	8,000	Mean
Morning......	4.3	4.0	3.4	3.0	2.4		3.4	3.4		3.4
Afternoon....										
Mean...	4.3	4.0	3.4	3.0	2.4		3.4	3.4		3.4

COMBINED.

	1,000	1,500	2,000	3,000	4,000	5,000	6,000	7,000	8,000	Mean
Morning......	4.6	3.9	3.6	3.2	3.0	2.8	3.1	3.0	3.0	3.4
Afternoon....	8.1	6.5	5.8	5.5	4.9	5.1				6.9
Mean...	5.6	4.4	4.0	3.5	3.2	3.0	3.1	3.0	3.0	3.6

INVERSIONS.

CLEAR.

Date.	400 feet. Rate.	Gradient.	800 feet. Rate.	Gradient.	1,000 feet. Rate.	Gradient.	1,200 feet. Rate.	Gradient.	1,500 feet. Rate.	Gradient.	2,000 feet. Rate.	Gradient.	2,500 feet. Rate.	Gradient.	3,000 feet. Rate.	Gradient.
May 12, a.m....													556	1.8		
14, a.m....			103	9.7	210	4.8										
25, a.m....			471	2.1												
31, a.m....					604	1.7	2,696	0.4								
June 1, a.m....			69	14.5	91	11.0										
1, a.m....			108	9.3												
14, a.m....							284	3.5	563	1.7	2,380	0.4				
15, a.m....	86	11.6			275	3.6	743	1.3								
21, a.m....			62	16.1			101	9.9	170	5.9	392	2.6	1,731	0.6		
22, a.m....	78	12.8	188	5.3			392	2.6	1,228	0.8						
23, a.m....	54	18.5	115	8.7			336	3.0								
24, a.m....			72	13.9			106	9.4	143	7.0	211	4.7	403	2.5		
25, a.m....			69	14.5			104	9.6	146	6.8	218	4.6	982	1.0		
26, a.m....			72	13.9			126	7.9								

Inversions—Clear—Continued.

Date.	400 feet.		800 feet.		1,000 feet.		1,200 feet.		1,500 feet.		2,000 feet.		2,500 feet.		3,000 feet.	
	Rate.	Gradient.	Rate.	Gradient.	Rate.	Gradient.	Rate.	Gradient.	Rate.	Gradient.	Rate.	Gradient.	Rate.	Gradient.	Rate.	Gradient.
July 3, a. m....			130	7.7	158	6.3			220	4.5	259	3.9	767	1.3		
3, a. m....			223	4.5												
4, a. m....			195	5.1			172	5.8	221	4.5	383	2.6	602	1.7		
10, a. m....	85	11.8	192	5.2			260	3.8	539	1.9	1,187	0.8				
11, a. m....	55	18.2	100	10.0	137	7.3	236	4.2								
Aug. 2, a. m....	149	6.7	236	4.2	546	1.8	778	1.3								
Sept. 1, a. m....			88	11.4			147	6.8								
2, a. m....			252	4.0			276	3.6	356	2.8						
27, a. m....	147	6.8	578	1.7												
Oct. 12, a. m....			266	3.7			433	2.3	857	1.2	4,240	0.2				
15, a. m....			838	1.2												
19, a. m....									3,472	0.4						
23, a. m....			234	4.3												
25, a. m....			120	8.3			167	6.0	334	3.0	636	1.6	1,677	0.6		
31, a. m....			111	9.0			509	2.0	1,658	0.6						

CLOUDY.

Date.	400 feet.		800 feet.		1,000 feet.		1,200 feet.		1,500 feet.		2,000 feet.		2,500 feet.		3,000 feet.	
	Rate.	Gradient.	Rate.	Gradient.	Rate.	Gradient.	Rate.	Gradient.	Rate.	Gradient.	Rate.	Gradient.	Rate.	Gradient.	Rate.	Gradient.
May 11, a. m....					404	2.5			1,503	0.6	1,187	0.8				
19, a. m....			819	1.2			691	1.4					1,106	0.9	6,876	0.2
24, 5 p. m....							509	2.0								
July 22, a. m....			546	1.8												
Aug. 14, a. m....			346	2.9			636	1.6	3,272	0.3						
Oct. 6, a. m....			182	5.5			582	1.7	1,460	0.7						
8, a. m....			970	1.0			1,044	1.0	988	1.0						
21, a. m....			2,970	0.3									1,570	0.6		

CAIRO, ILL.

TEMPERATURE.

At Cairo there were considered 39 ascensions and 116 observations at altitudes of 1,000 feet or more, and the greatest altitude attained was 5,577 feet.

The mean gradient was 6° for each 1,000 feet. It decreased quite rapidly up to 3,000 feet and very slowly thereafter. The value up to 1,000 feet is particularly large, being 9.7° per thousand feet, while up to 1,500 feet there is a sharp fall to 6.6°.

The morning gradients were steadily and quite uniformly less than the afternoon ones, except up to 1,000 feet, where the morning value reached 10.5° per thousand feet, 2.5° greater than the afternoon one. The averages for all elevations differed but little.

The clear and cloudy weather gradients did not maintain a steady relation toward each other. Up to 2,000 and 5,000 feet the former were the greater, while up to 3,000 feet there was a marked change in the opposite direction. Above 4,000 feet there was very little difference. Cloud effects were as a rule quite marked. There was usually a decided decrease in the rate of temperature fall, and in a number of cases the fall was changed into a rise. After the kite emerged from the clouds the temperature change would be slower than it had been before the clouds were encountered. On July 13 the temperature fall was only 4.5° while the kite was rising from 2,700 to 5,000 feet. From 3,750 to 5,000 feet the fall was but 0.5°. On July 28 there was rise of 3° while the kite rose from 3,150 to 3,450 feet, followed by a further rise of 1° up to 4,000 feet, where the kite emerged from the clouds. Above 4,000 feet a slow fall commenced. On September 4, while a thunderstorm was in progress, there was a rise in the temperature of 2° just as the rain began, followed in a few minutes by a fall of 6° while the kite was rising from 2,100 to 2,800 feet.

Neither wind direction nor velocity appeared to exert any influence upon the temperature gradient.

WINDS.

Wind directions did not vary materially from the usual rule. Neither was the amount of deflection of the kite marked except in one instance, when it varied from 90° to 135° just after the termination of a shower. There were five cases of deflection toward the left, varying in amount from 30° to 90°. In two of these cases a thunderstorm followed shortly.

RELATIVE HUMIDITY AND VAPOR PRESSURE.

The relative humidities, both above and at the surface, were almost exactly alike, the extreme difference above being but 5 per cent and that at the surface 8 per cent. Except at 5,000 feet the difference between the surface and upper-air humidities did not exceed 1 per cent.

Cloud effects were quite noticeable both while the kite was enveloped in clouds and while rain conditions were present. A rise usually resulted, followed by a fall, as the cloud or rain conditions disappeared. On July 28 there was a fall in the humidity from 90 to 42 per cent as the kite emerged from the clouds.

The vapor-pressure results were very similar to those obtained at Cincinnati and Springfield— stations also located in the great Mississippi Basin.

DIMINUTION OF VAPOR PRESSURE WITH ALTITUDE.

VALUE OF $\frac{p}{p_0}$ AT EACH RESPECTIVE 1,000 FEET OF ALTITUDE.

1,500 feet.	2,000 feet.	3,000 feet.	4,000 feet.	5,000 feet.
0.71	0.69	0.63	0.54	0.54

TEMPERATURE.
CLEAR.

Date.	1,000 feet.		1,500 feet.		2,000 feet.		3,000 feet.		4,000 feet.		5,000 feet.		Mean.	
	Rate.	Gradient.	Rate.	Gradient.	Rate.	Gradient.	Rate.	Gradient.	Rate.	Gradient.	Rate.	Gradient.	Rate.	Gradient.
May 15, p. m					216	4.6	198	5.1	205	4.9	215	4.7		
18, a. m			164	6.1	189	5.3	212	4.7	207	4.8	210	4.8		
27, p. m			111	9.0	152	6.5	196	5.1	227	4.4				
29, p. m			129	7.8	170	5.9								
30, a. m			121	8.3	113	8.8								
June 5, p. m			157	6.4	137	7.3								
13, p. m			218	4.6										
19, p. m			125	8.0	133	7.5	155	6.5						
24, p. m			114	8.8	143	7.0	163	6.1	157	6.4				
July 9, a. m			158	6.3	253	3.8	274	3.6	276	3.7	272	3.7		
10, a. m			151	6.6	202	5.0	341	2.9	236	4.2				
11, a. m							669	1.6	240	4.2				
12, a. m			133	7.5	148	6.8	158	6.3						
19, a. m	118	8.5	161	6.2										
Aug. 7, p. m	167	6.0												
Sept. 3, p. m			178	5.6	200	5.0								
5, a. m			188	5.3	186	5.4	333	3.0	242	4.1				
7, p. m	100	10.0												
10, a. m	71	14.1												
11, a. m			97	10.3	106	9.3								
Oct. 16, p. m			214	4.7	170	5.9	177	5.6	179	5.6				
23, p. m			131	7.6	174	5.7	283	3.5						
24, a. m			121	8.3	149	6.7	273	3.7	267	3.7				
30, p. m			125	8.0	152	6.6								

CLOUDY.

Date.	1,000 feet.		1,500 feet.		2,000 feet.		3,000 feet.		4,000 feet.		5,000 feet.		Mean.	
May 19, a. m	74	13.5	109	9.2	133	7.5	143	7.0	165	6.1				
20, a. m	167	6.0	175	5.7	163	6.1	183	5.5	200	5.0				
June 11, p. m					139	7.2	152	6.6	187	5.3				
17, p. m			130	7.7	143	7.0	163	6.1						
25, a. m			170	5.9	241	4.1								
25, a. m			118	8.5	125	8.0	145	6.9	178	5.6				
26, a. m			350	2.0	278	3.6	242	4.1	267	3.7				
July 4, a. m			194	5.2										
8, a. m			177	5.6	154	6.5								
13, a. m			876	1.1	563	1.8	372	2.7	310	3.2	258	3.9		
20, a. m			141	7.1	182	5.5	196	5.4	222	4.5				
28, a. m			177	5.6	230	4.3	232	4.3						
28, p. m									250	4.0				
Sept. 4, p. m			135	7.4	157	6.4								
21, p. m			255	3.9										
22, a. m			126	7.9	159	6.8	167	6.0	178	5.6				

Temperature—Continued.

SUMMARY.

CLEAR.

Date.	1,000 feet.		1,500 feet.		2,000 feet.		3,000 feet.		4,000 feet.		5,000 feet.		Mean.	
	Rate.	Gradient.	Rate.	Gradient.	Rate.	Gradient.	Rate.	Gradient.	Rate.	Gradient.	Rate.	Gradient.	Rate.	Gradient.
Morning........	....	11.3	....	5.6	....	6.4	....	3.7	....	4.1	....	4.2	....	5.9
Afternoon......	....	8.0	....	6.9	....	6.2	....	5.3	....	5.3	....	4.7	....	6.1
Mean	....	9.8	....	6.0	....	6.3	....	4.4	....	4.6	....	4.4	....	5.9

CLOUDY.

Morning........	....	9.8	....	7.2	....	5.1	....	4.9	....	4.7	....	3.9	....	5.9
Afternoon......	....		....	7.0	....	7.2	....	6.9	....	5.0	....		....	6.5
Mean	....	9.8	....	7.1	....	5.7	....	5.5	....	4.8	....	3.9	....	6.5

COMBINED.

Morning........	....	10.5	....	6.4	....	5.7	....	4.3	....	4.4	....	4.1	....	5.9
Afternoon......	....	8.0	....	7.0	....	6.5	....	5.8	....	5.2	....	4.7	....	6.2
Mean	....	9.7	....	6.6	....	6.0	....	4.9	....	4.7	....	4.3	....	6.0

CINCINNATI, OHIO.

TEMPERATURE.

At Cincinnati there were 38 ascensions and 81 observations at altitudes of 1,000 feet or more, and the greatest altitude reached was 6,313 feet.

The mean decrease in temperature with increase of altitude was 6.6° for each 1,000 feet, the highest gradient found at any station. The abnormally large gradient of 13° per thousand feet occurs up to 1,000 feet, but as there were but three observations at this altitude it is very probable that a greater number would have resulted in a reduced gradient. Up to 1,500 feet the gradient is less than one-half as large.

The rate of temperature diminution decreased, as usual, with increase of altitude, except up to 2,000 feet, but it is perhaps true that the increase at this elevation is more apparent than real, as the gradient up to 1,500 feet is comparatively small and is the mean of but seven observations. At 1,500 and from 3,000 up to 5,000 feet the morning and afternoon gradients were nearly equal, but up to 1,000, 2,000, and 6,000 feet there is considerable variation, the afternoon ones being the greater.

While the mean clear and mean cloudy weather gradients were exactly equal, they varied greatly and irregularly at the different elevations. At some elevations the clear ones were the greater and at others the cloudy ones. The remarkable gradient of 14.7° per thousand feet was noted in the early afternoon of June 8, during cloudy weather.

In the few instances in which the kite was in proximity to the clouds there did not appear to have been any variations in the temperatures that could not be accounted for by difference of elevation.

WINDS.

Wind directions adhered to the usual rule of deflection toward the right, but not to any decided degree, the maximum deflection having been about 90°, and that in one or two cases only. There were four cases of deflection in the opposite direction of from 20° to 40°, and three of them were followed by rain in one or two hours.

RELATIVE HUMIDITY AND VAPOR PRESSURE.

The relative humidity results above agreed very closely with those at the surface, the greatest difference, 7 per cent, being found at 6,000 feet. With this exception the greatest difference was 3 per cent.

The approach of clouds or rain had the usual effect of producing a rise in the humidity, which was more marked in the case of rain. Vapor-pressure results follow below. There is an excellent agreement with the values obtained at Springfield, Ill., a neighboring station.

DIMINUTION OF VAPOR PRESSURE WITH ALTITUDE.

VALUE OF $\frac{p}{p'}$ AT EACH RESPECTIVE 1,000 FEET OF ALTITUDE.

1,500 feet.	2,000 feet.	3,000 feet.	4,000 feet.	5,000 feet.	6,000 feet.
0.73	0.67	0.57	0.51	0.49	0.45

TEMPERATURE.

CLEAR.

Date.	1,000 feet. Rate.	Gradient.	1,500 feet. Rate.	Gradient	2,000 feet. Rate.	Gradient.	3,000 feet. Rate.	Gradient.	4,000 feet. Rate.	Gradient.	5,000 feet. Rate.	Gradient.	6,000 feet. Rate.	Gradient.	Mean. Rate.	Gradient.
May 8, a. m.									457	2.2	526	1.9	342	2.9		
15, p. m.							202	5.0								
29, p. m.					168	6.0										
June 3, p. m.			106	9.4												
4, a. m.									364	2.7	217	4.6				
4, p. m.					99	10.0										
4, p. m.					82	12.2										
15, p. m.			94	10.6	177	5.6										
19, a. m.					83	12.0										
19, p. m.	77	13.0					120	8.3								
24, a. m.							104	9.6								
July 3, a. m.					148	6.8										
9, a. m.			500	2.0												
9, p. m.					154	6.5	340	2.9								
10, a. m.					135	7.4	139	7.2								
19, a. m.			109	9.2												
19, p. m.					96	10.4										
28, p. m.			256	3.9	173	5.8										
29, a. m.							165	6.1	170	5.9	199	5.0				
Aug. 3, a. m.					164	6.1	176	5.7	211	4.7						
3, a. m.									173	5.8	208	4.8				
12, a. m.					174	5.7										
29, a. m.	86	11.6														
Sept. 6, a. m.							353	2.8	338	3.0						
6, a. m.			131	7.6	155	6.5	158	6.3	170	5.9						
7, a. m.							171	5.8	173	5.8						
11, a. m.					1,046	1.0										
12, a. m.							230	4.3								
14, p. m.					117	8.5										
18, a. m.					171	5.8	177	5.6	216	4.6	242	4.1				
24, p. m.					100	10.0	149	6.7								
27, p. m.					125	8.0										

CLOUDY.

Date.	1,000 feet. Rate.	Gradient.	1,500 feet. Rate.	Gradient	2,000 feet. Rate.	Gradient.	3,000 feet. Rate.	Gradient.	4,000 feet. Rate.	Gradient.	5,000 feet. Rate.	Gradient.	6,000 feet. Rate.	Gradient.	Mean. Rate.	Gradient.
May 10, a. m.									67	14.9						
10, p. m.											189	5.3	193	5.2		
11, a. m.							158	6.3								
11, p. m.									138	7.2						
28, a. m.											267	6.7	222	4.5		
28, p. m.											216	4.6				
June 8, p. m.	68	14.7														
11, p. m.									244	4.1						
July 18, a. m.					186	5.4										
20, a. m.					131	7.6	154	6.5	176	6.0	179	5.6				
25, p. m.					148	6.8	174	5.7								
Aug. 3, p. m.									206	4.9						
24, a. m.							171	5.8	178	5.6						
Sept. 4, a. m.			819	1.1	657	1.5	460	2.1								
5, a. m.					962	1.0										
26, a. m.					122	8.2										
Oct. 1, p. m.							136	7.4	170	5.9						

Temperature—Continued.

SUMMARY.

CLEAR.

Date.	1,000 feet.		1,500 feet.		2,000 feet.		3,000 feet.		4,000 feet.		5,000 feet.		6,000 feet.		Mean.	
	Rate.	Gradient.	Rate.	Gradient.	Rate.	Gradient.	Rate.	Gradient.	Rate.	Gradient.	Rate.	Gradient.	Rate.	Gradient.	Rate.	Gradient.
Morning		11.6		7.4		5.9		5.9		4.5		4.1		2.9		6.0
Afternoon		13.0		6.6		9.0		5.7								8.0
Mean		12.3		7.1		7.5		5.9		4.5		4.1		2.9		6.3

CLOUDY.

	Rate.	Gradient.	Rate.	Gradient.	Rate.	Gradient.	Rate.	Gradient.	Rate.	Gradient.	Rate.	Gradient.	Rate.	Gradient.	Rate.	Gradient.
Morning				1.1		4.7		5.2		8.8		6.2		4.5		5.1
Afternoon		14.7				6.8		6.6		5.5		5.0		5.2		7.3
Mean		14.7		1.1		5.1		5.6		6.9		5.6		4.8		6.3

COMBINED.

	Rate.	Gradient.	Rate.	Gradient.	Rate.	Gradient.	Rate.	Gradient.	Rate.	Gradient.	Rate.	Gradient.	Rate.	Gradient.	Rate.	Gradient.
Morning		11.6		6.3		5.5		5.7		5.6		4.7		3.7		6.2
Afternoon		13.8		6.6		8.8		6.0		5.5		5.0		5.2		7.3
Mean		13.0		6.3		6.9		5.8		5.6		4.7		4.2		6.6

FORT SMITH, ARK.

TEMPERATURE.

Owing to the absence of sufficient wind to fly the kite, there were only 19 ascensions and 58 observations made at Fort Smith at altitudes of 1,000 feet or more. Of these 6 ascensions and 19 observations were made during June; 4 ascensions and 16 observations during July; 2 ascensions and 4 observations during August, and 7 ascensions and 19 observations during September. The highest altitude attained was 5,431 feet.

The average hourly wind velocity at the surface was 5.6 miles from May to October, inclusive, with a maximum average of 6.3 miles in May and a minimum of 4.5 miles in August.

The mean decrease in temperature with increase of altitude was 6.1° for each 1,000 feet. The gradient decreased slowly up to 2,000 feet, and quite rapidly thereafter, a departure from the conditions generally prevailing elsewhere.

The observations were equally divided between the mornings and afternoons, and the means were exactly equal. The gradients also differed but little up to the different altitudes, except at 1,000 feet, where the morning one was 2.6° larger.

The clear and cloudy weather gradients were exactly alike up to 2,000 feet; up to 1,500 feet the cloudy one was 1.2° larger than the clear one, while up to the remaining altitudes the clear weather ones were the greater.

There were but two instances in which the kite was enveloped in clouds, and both were apparently without effect upon the temperature. Wind directions and velocities were without influence upon the gradients. It is of interest to note, however, in connection with the winds, that there were no ascensions made with northerly winds, and that the greater portion were made with southwesterly winds, except in September, when they were mostly southeasterly.

WINDS.

The wind directions conformed to the usual rule, although the deflections were slight, except in one instance when it amounted to 90°. But one case of deflection toward the left was noticed. It was quite well marked, gradually increasing with increase of altitude from 25° to 60°, and was followed within an hour by light showers of very brief duration.

RELATIVE HUMIDITY AND VAPOR PRESSURE.

The relative humidities were considerably greater above than at the surface, the greatest difference—19 per cent—occurring at 3,000 feet and the least—8 per cent—at 4,000 feet. The mean difference was 12 per cent.

Cloud effects were, as usual, confined to an increase in the humidity as the kite entered the clouds, followed by a corresponding fall as the kite emerged.

Vapor pressure results as follows were determined. They do not correspond very well with any others found in the West except those at Omaha:

DIMINUTION OF VAPOR PRESSURE WITH ALTITUDE.

VALUE OF $\frac{p}{p^o}$ AT EACH RESPECTIVE 1,000 FEET OF ALTITUDE.

1,500 feet.	2,000 feet.	3,000 feet.	4,000 feet.
0.82	0.79	0.74	0.65

TEMPERATURE.
CLEAR.

Date.	1,000 feet. Rate.	1,000 feet. Gradient.	1,500 feet. Rate.	1,500 feet. Gradient.	2,000 feet. Rate.	2,000 feet. Gradient.	3,000 feet. Rate.	3,000 feet. Gradient.	4,000 feet. Rate.	4,000 feet. Gradient.	Mean. Rate.	Mean. Gradient.
June 26, a. m.	129	7.8	142	7.0	154	6.5						
26, p. m.							144	6.9				
July 19, a. m.	104	9.6			143	7.0	147	6.8				
19, p. m.			192	5.2	190	5.3	145	6.9				
28, p. m.			146	6.8	141	7.1	150	6.3				
Aug. 5, a. m.	128	7.8										
Sept. 3, p. m.			144	6.9	108	9.3	194	5.2				
4, a. m.			263	3.8	189	5.3						
5, a. m.			248	4.0	205	4.9	192	5.2				
5, p. m.			158	6.3								
15, a. m.			91	11.0	115	8.7						
15, p. m.							221	4.5				
16, a. m.	142	7.0	160	6.2	159	6.3						

CLOUDY.

Date.	1,000 feet. Rate.	1,000 feet. Gradient.	1,500 feet. Rate.	1,500 feet. Gradient.	2,000 feet. Rate.	2,000 feet. Gradient.	3,000 feet. Rate.	3,000 feet. Gradient.	4,000 feet. Rate.	4,000 feet. Gradient.	Mean. Rate.	Mean. Gradient.
June 11, a. m.			119	8.4	113	8.8						
16, p. m.			122	8.2								
20, a. m.	120	8.3	102	9.7	119	8.4						
24, a. m.	101	9.9										
24, p. m.			132	7.6	129	7.8						
25, a. m.	128	7.8	136	7.4	161	6.2	211	4.7	274	3.6		
25, a. m.									244	4.1		
July 18, a. m.	127	7.9	141	7.1	154	6.5						
18, p. m.					146	6.8						
30, p. m.	492	2.0	179	5.6	246	4.1						
Aug. 31, p. m.	278	3.6	147	6.8	184	5.4						
Sept. 17, a. m.	145	6.9										
29, a. m.	133	7.5	136	7.4								
29, p. m					166	6.0						

SUMMARY.
CLEAR.

	1,000 feet.	1,500 feet.	2,000 feet.	3,000 feet.	4,000 feet.	Mean.
Morning	8.0	6.4	6.4	6.0		6.7
Afternoon		6.3	7.2	6.0		6.5
Mean	8.0	6.4	6.7	6.0		6.8

CLOUDY.

	1,000 feet.	1,500 feet.	2,000 feet.	3,000 feet.	4,000 feet.	Mean.
Morning	7.7	7.6	7.2	4.7		6.2
Afternoon	5.2	7.6	6.4		3.8	6.4
Mean	6.7	7.6	6.7	4.7	3.8	5.9

COMBINED.

	1,000 feet.	1,500 feet.	2,000 feet.	3,000 feet.	4,000 feet.	Mean.
Morning	7.8	6.9	6.7	5.6	3.8	6.2
Afternoon	5.2	7.1	6.7	6.0		6.2
Mean	7.2	7.0	6.7	5.8	3.8	6.2

KNOXVILLE, TENN.

TEMPERATURE.

At Knoxville there were only 19 ascensions and 34 observations at altitudes of 1,000 feet or more, and the highest altitude attained was 4,482 feet. On May 5, before the regular observations were commenced, an altitude of 6,058 feet was reached.

The mean decrease in temperature with increase of altitude was 6.3° for each 1,000 feet. Owing, most likely, to the great scarcity of observations, the rate of change of the temperature varied irregularly as the elevation increased, instead of decreasing steadily. Except up to 4,000 feet, the morning gradients were greater than the afternoon ones, particularly at 1,500 and 2,000 feet. The average morning result was 0.6° greater than the afternoon one.

As there were but three observations made during cloudy weather, the results given above are practically clear-weather results.

There were but two ascensions in which the kite came into contact with clouds. In one the clouds appeared to be without effect, while in the other a slight retardation in the rate of temperature fall was noticed.

Neither wind direction nor velocity appeared to have any bearing upon the results.

WINDS.

Wind directions conformed to the usual rule with but one or two exceptions. On one day there was a divergence of the kite toward the left of about 20°, and a thunderstorm occurred several hours later.

RELATIVE HUMIDITY AND VAPOR PRESSURE.

The relative humidity was higher above than at the surface, except at 4,000 feet. The greatest difference was 10 per cent at 2,000 feet.

The vapor pressure results follow: Up to 3,000 feet they agree very well with others from the Ohio Valley northward, and are not greatly dissimilar to those found at Washington. At 4,000 feet, where there were but two observations, the results appear to be abnormally high.

DIMINUTION OF VAPOR PRESSURE WITH ALTITUDE.

VALUE OF $\frac{p}{p'}$ AT EACH RESPECTIVE 1,000 FEET OF ALTITUDE.

1,500 feet.	2,000 feet.	3,000 feet.	4,000 feet.
0.83	0.73	0.66	0.76

TEMPERATURE.

CLEAR.

Date.	1,000 feet.		1,500 feet.		2,000 feet.		3,000 feet.		4,000 feet.		Mean.	
	Rate.	Gradient.	Rate.	Gradient.	Rate.	Gradient.	Rate.	Gradient.	Rate.	Gradient.	Rate.	Gradient.
May 10, p. m			134	7.5	101	9.9	98	10.0				
11, p. m	158	6.3										
12, a. m			146	6.8								
16, a. m	176	5.7					337	3.0				
25, p. m												
26, p. m			266	3.8								
30, p. m			103	9.7	101	9.9						
June 3, p. m			263	3.8	241	4.1						
3, p. m	400	2.5	375	2.7	267	3.7						
4, p. m	91	11.0										
21, p. m					161	6.2						
25, a. m	110	9.1										
July 10, p. m							152	6.6				
11, a. m					153	6.5	153	6.5				
11, p. m	100	10.0	114	8.8	140	7.1	342	4.1				
12, a. m			150	6.7	187	5.3						
12, p. m												
Aug. 3, p. m	101	9.9										
Sept. 23, a. m							221	4.5	222	4.5		

CLOUDY.

June 17, p. m	81	12.3										
24, p. m							238	4.2	183	5.5		
July 12, p. m							235	4.3				

Temperature—Continued.
SUMMARY.
CLEAR.

Date.	1,000 feet.		1,500 feet.		2,000 feet.		3,000 feet.		4,000 feet.		Mean.	
	Rate.	Gradient.	Rate.	Gradient.	Rate.	Gradient.	Rate.	Gradient.	Rate.	Gradient.	Rate.	Gradient.
Morning	....	9.1	....	8.2	....	8.2	....	4.5	....	4.5	....	6.9
Afternoon	....	7.0	....	5.6	....	6.0	....	8.3	....	....	....	6.9
Mean	....	7.8	....	6.2	....	6.6	....	5.8	....	4.5	....	6.2

CLOUDY.

Morning												
Afternoon	....	12.3	....		....		....	4.2	....	5.5	....	7.3
Mean	....	12.3	....		....		....	4.2	....	5.5	....	7.3

COMBINED.

Morning	....	9.1	....	8.2	....	8.2	....	4.5	....	4.5	....	6.9
Afternoon	....	8.2	....	5.6	....	6.0	....	6.3	....	5.3	....	6.3
Mean	....	8.4	....	6.2	....	6.6	....	5.4	....	5.0	....	6.3

MEMPHIS, TENN.
TEMPERATURE.

At Memphis there were 37 ascensions and 117 observations at altitudes of 1,000 feet or more, and the greatest height reached was 5,243 feet.

The mean decrease in temperature with increase of altitude was 5.1° per thousand feet, and there was a steady decrease in the rate up to the highest altitude. The decrease was quite rapid up to 3,000 feet and very slow thereafter.

The morning gradients were considerably less than the afternoon ones except up to 1,000 and 5,000 feet. Up to the former elevation there is a difference of 1.5° per thousand feet, the morning gradient reaching the high value of 8.7° per thousand feet. Up to 5,000 feet the difference in favor of the morning gradient was trifling.

By way of exception to the usual rule, the cloudy weather gradients were greater than the clear weather ones up as far as 1,500 feet, the difference being 1.4° per thousand feet up to the latter elevation. Up to the higher elevations the usual rule holds, but the differences were comparatively small.

Cloud effects, while not at all decided, were quite frequent and constant. There was usually a suspension of the temperature change as the kite came into contact with the clouds, followed by a slower rise or fall than had obtained before the clouds were encountered.

Wind velocities were without effect. The directions, however, appeared to have some bearing upon the gradients. It was noticed that with winds from the northwest to east the temperature decreased more rapidly than with those from any other direction. The greatest gradients were found with north to east winds and the least with south to west ones.

WINDS.

The usual rule of deflection toward the right obtained except in one or two insignificant instances.

RELATIVE HUMIDITY AND VAPOR PRESSURE.

The relative humidities above 1,500 feet did not differ greatly, the extreme range being 15 per cent. They were greatest at 2,000 and 3,000 feet, and least at 5,000 feet. At 3,000 feet they were greater than the surface humidities, while at 4,000 and 5,000 feet the two were almost exactly alike.

Clouds and the near approach of rain caused a rise in the humidity, which in the former case was followed by a fall as soon as the kite was removed from cloud influence.

The vapor pressure results are given below. At the lower elevations they agree quite well with others in the Upper Mississippi Valley. The nearest approach to the Memphis values was found at Lansing, Mich.

DIMINUTION OF VAPOR PRESSURE WITH ALTITUDE.

VALUE OF $\frac{p}{p^6}$ AT EACH RESPECTIVE 1,000 FEET OF ALTITUDE.

1,500 feet.	2,000 feet.	3,000 feet.	4,000 feet.	5,000 feet.
0.88	0.86	0.76	0.61	0.54

TEMPERATURE.

CLEAR.

Date.	1,000 feet.		1,500 feet.		2,000 feet.		3,000 feet.		4,000 feet.		5,000 feet.		Mean.	
	Rate.	Gradient.	Rate.	Gradient.	Rate.	Gradient.	Rate.	Gradient.	Rate.	Gradient.	Rate.	Gradient.	Rate.	Gradient.
June 24, p. m	213	4.7	191	5.2	390	2.6								
24, p. m					248	4.0	196	5.1						
24, p. m			730	1.4	571	1.8								
26, a. m	101	9.9	131	7.6	119	8.4								
26, p. m			111	9.0										
July 10, a. m							760	1.3	267	3.7	227	4.4		
10, p. m			151	6.6	158	6.3	190	5.3						
11, a. m									312	3.2	311	3.2		
11, p. m	1,009	1.0	750	1.3	499	2.0	307	3.3						
12, p. m	173	5.8												
29, a. m					1,000	1.0	427	2.3	221	4.5				
30, a. m							284	3.5	244	4.1				
Sept. 4, a. m			172	5.8	215	4.7	261	3.8						
5, a. m							246	4.1	246	4.1				
5, p. m					133	7.5	168	6.0	174	5.7				
5, p. m									182	5.5				
6, a. m	116	8.6												
8, a. m	77	13.0												
14, a. m			200	5.0	166	6.0								
14, p. m			153	6.5										
14, p. m			141	7.1										
Oct. 3, a. m					192	5.2	219	4.6	222	4.5				
3, p. m			102	9.8	136	7.4								
11, p. m			136	7.4	141	7.1	162	6.2						
16, a. m					350	2.9	321	3.1	285	3.1				
16, p. m	101	9.9	123	8.1	133	7.5	170	5.9						
18, a. m							300	3.3	223	4.5				
22, a. m	163	6.1	102	9.7	222	4.5								
24, a. m							375	2.7	513	1.9				
24, p. m	107	9.3	130	7.7	145	6.9	207	4.8	259	3.9	293	3.4		
26, a. m			194	5.2	173	5.8	250	4.0						
26, p. m	117	8.5												

CLOUDY.

Date.	1,000 feet.		1,500 feet.		2,000 feet.		3,000 feet.		4,000 feet.		5,000 feet.		Mean.	
	Rate.	Gradient.	Rate.	Gradient.	Rate.	Gradient.	Rate.	Gradient.	Rate.	Gradient.	Rate.	Gradient.	Rate.	Gradient.
June 25, a. m							419	2.4	335	5.0				
27, a. m							545	1.8	398	2.5	314	3.2		
27, a. m					136	7.4	139	7.2	225	4.4				
July 8, a. m					250	4.0	294	3.4						
9, a. m							283	3.5	308	3.2				
9, a. m							273	3.7						
13, a. m					367	2.7								
17, a. m					554	1.8								
19, a. m			319	3.1										
19, p. m					304	3.3								
20, a. m							509	2.0	244	4.1				
20, a. m									225	4.4				
22, a. m	178	5.6												
25, a. m	174	5.7												
Aug. 26, p. m	271	3.7												
30, a. m	100	10.0	177	5.6	201	5.0								
30, a. m			84	11.9										
31, a. m	83	12.0												
Sept. 7, p. m			91	11.0	132	7.6								
9, p. m	85	11.8	126	7.8										
10, a. m					151	6.6								
12, a. m									992	1.0				
12, a. m							294	3.4	334	3.0				
12, p. m			128	7.8	158	6.3								
22, a. m							375	2.7	364	2.7	294	3.4		

Temperature—Continued.

SUMMARY.

CLEAR.

Date.	1,000 feet.		1,500 feet.		2,000 feet.		2,000 feet.		4,000 feet.		5,000 feet.		Mean.	
	Rate.	Gradient.	Rate.	Gradient.	Rate.	Gradient.	Rate.	Gradient.	Rate.	Gradient.	Rate.	Gradient.	Rate.	Grndient.
Morning..........	....	9.4	....	6.6	....	4.6	....	3.3	....	3.7	....	3.8	....	5.2
Afternoon	....	6.5	....	6.4	....	5.4	....	5.2	....	5.0	....	3.4	....	5.3
Mean	....	7.7	....	6.5	....	5.1	....	4.1	....	4.1	....	3.7	....	5.2

CLOUDY.

Morning.......	....	7.8	....	4.4	....	4.6	....	3.3	....	3.4	....	3.3	....	4.5
Afternoon	....	8.5	....	9.6	....	5.7	....		....		....		....	7.9
Mean	....	8.1	....	7.9	....	5.0	....	3.3	....	3.4	....	3.3	....	5.2

COMBINED.

Morning..........	....	8.7	....	6.0	....	4.6	....	3.3	....	3.6	....	3.6	....	5.0
Afternoon	....	7.2	....	7.2	....	5.4	....	5.2	....	5.0	....	3.4	....	5.6
Mean	....	7.8	....	6.8	....	5.0	....	3.8	....	3.7	....	3.5	....	5.1

SPRINGFIELD, ILL.

TEMPERATURE.

Of the Springfield series of observations 174 were considered, having been made during 46 ascensions at altitudes of from 1,000 to 6,000 feet.

The mean rate of decrease of temperature with increase of altitude was found to be 4.9° for each 1,000 feet, the lowest gradient in the Mississippi Valley basin. The greatest gradient, 7.6° for each thousand feet, was found at an elevation of 1,000 feet, and there is a steady decrease up to the highest limits reached. The morning and afternoon curves differ but slightly above 2,000 feet; from the ground up to 2,000 feet the difference varied from 0.7° to 1.5°, being greatest up to 1,500 feet. If all altitudes are considered together, the morning, afternoon, and mean results are almost exactly alike.

The clear and cloudy weather gradients were also quite similar, with values of 5.1° and 4.6° per thousand feet, respectively. The only marked difference was found up to 1,000 feet, where it was 2.4° per thousand feet.

The principal effect of clouds upon the temperature was a diminution of the rate of fall, at times amounting to a complete suspension. On October 29 there was a rise of 5° from 2,100 to 3,700 feet while the kite was in the clouds, and, in consequence, a considerable decrease in the gradient. The opposite effect was noticed on July 18, when there was a fall of 4° while the kite was in the clouds, followed by a rise of 2.5° when it emerged. In some instances the clouds appeared to have no effect.

Wind effects were confined to a decrease in the gradient with increase of velocity, although the two were not at all proportional.

WINDS.

The wind directions conformed to the rule of general agreement which has uniformly obtained. There was the same deflection of the kite toward the right in the upper air; it increased with increase of altitude, but rarely amounted to 90 degrees or more. In one case there was a deflection of 90 degrees toward the left during the ascension, caused by the shifting of the surface wind, the direction of the upper remaining the same, and a severe thunderstorm took place about eight hours after the change occurred.

RELATIVE HUMIDITY AND VAPOR PRESSURE.

The relative humidity was decidedly less above than at the surface, but there was comparatively little change above 1,500 or 2,000 feet, except when clouds were present or when rain had fallen a few hours before or after. When there were clouds there would be a marked rise, amounting at times to total saturation, as long as the kite was in the clouds, a fall taking place as soon as it emerged.

When rain had fallen shortly before or after the ascension the humidity would be high, usually higher than at the surface, and would remain so for hours, although the surface humidity was falling rapidly. In clear weather the upper air humidity was usually less than that at the surface, sometimes approaching zero, while during cloudy or rainy weather the upper would be the higher, but not with so great differences as obtained under reverse conditions.

The vapor pressure percentages follow below:

DIMINUTION OF VAPOR PRESSURE WITH ALTITUDE.

VALUE OF $\frac{p}{p_0}$ AT EACH RESPECTIVE 1,000 FEET OF ALTITUDE.

1,500 feet.	2,000 feet.	3,000 feet.	4,000 feet.	5,000 feet.	6,000 feet.
0.74	0.70	0.63	0.52	0.48	0.49

TEMPERATURE

CLEAR.

Date.	1,000 feet.		1,500 feet.		2,000 feet.		3,000 feet.		4,000 feet.		5,000 feet.		6,000 feet.		Mean.	
	Rate.	Gradient.	Rate.	Gradient.	Rate.	Gradient.	Rate.	Gradient.	Rate.	Gradient.	Rate.	Gradient.	Rate.	Gradient.	Rate.	Gradient.
June 10, a. m...			348	2.9	231	4.3	224	4.3								
19, a. m...					229	4.4										
19, a. m...					140	7.1										
21, a. m...	110	9.1														
24, a. m...			316	3.2	198	5.1										
24, p. m...			204	4.9												
29, a. m...					606	1.7	266	3.8								
29, p. m...			158	6.3	141	7.1										
July 1, p. m...	85	11.0	119	8.4	115	8.7										
2, a. m...			175	5.7	182	5.5	187	5.3								
9, a. m...					806	1.2										
10, a. m...	104	9.6	131	7.6	122	8.2										
13, p. m...	95	10.5	208	4.8	182	5.5										
14, p. m...	81	12.3	143	7.0												
20, a. m...			153	6.5												
27, p. m...			109	9.2												
Aug. 15, p. m...					257	3.9	342	2.9	302	3.3	300	3.3				
21, a. m...			148	6.8	176	5.7	196	5.1								
21, p. m...			134	7.5	174	5.7			235	5.4	213	4.7				
23, p. m...	102	6.2	150	6.7	177	5.6	241	4.1								
23, p. m...			430	2.3	314	3.2										
30, p. m...			204	4.9	246	4.1										
27, p. m...	119	8.4	110	9.1	157	6.4										
29, a. m...	161	6.2	463	2.2	352	2.8	250	4.0								
30, p. m...	253	4.0	193	5.2	162	6.2										
Sept. 29, p. m...			152	6.6	176	5.7	202	5.0	188	5.3	207	4.8				
29, p. m...									183	5.5						
Oct. 2, p. m...			195	5.2	174	5.7	187	5.3								
23, p. m...							610	1.6	457	2.2	538	1.9				
26, p. m...					136	7.4	181	5.5	176	5.7	214	4.7				
27, p. m...			156	6.4	142	7.0										
27, p. m...							264	3.8	331	3.0						
27, p. m...			247	4.0	224	4.5	270	2.7								
28, p. m...			140	7.1												
28, p. m...			136	7.4	175	5.7	259	3.9	232	4.3	238	4.2	270	3.7		
28, p. m...											302	3.3				
31, p. m...			150	6.7	276	3.6										
31, p. m...			741	1.3	702	1.4										

CLOUDY.

Date.	1,000 feet.		1,500 feet.		2,000 feet.		3,000 feet.		4,000 feet.		5,000 feet.		6,000 feet.		Mean.	
June 12, a. m...			132	7.6	154	6.5	154	6.5								
15, a. m...	265	3.8	635	1.6	434	2.3										
25, a. m...	316	3.2	252	4.0	254	3.9			250	4.0						
27, a. m...	124	8.1			181	5.5	210	4.8								
27, p. m...	105	9.5			123	8.1	181	5.5								
July 17, p. m...			105	9.5	310	9.1										
17, p. m...			274	3.6												
18, a. m...			131	7.6	165	6.1	163	6.1								
18, p. m...					127	7.9	132	7.6								
28, a. m...			330	3.0	494	2.0										

Temperature—Cloudy—Continued.

Date.	1,000 feet.		1,500 feet.		2,000 feet.		3,000 feet.		4,000 feet.		5,000 feet.		6,000 feet.		Mean.	
	Rate.	Gradient.	Rate.	Gradient.	Rate.	Gradient.	Rate.	Gradient.	Rate.	Gradient.	Rate.	Gradient.	Rate.	Gradient.	Rate.	Gradient.
Aug. 2, a. m....	128	7.8	128	7.8												
3, p. m....			95	10.5	115	8.7	163	6.1								
11, p. m....	215	4.7	213	4.7	203	4.9	211	3.2								
16, p. m....			152	6.6	196	5.1	205	4.9	241	4.1						
20, p. m....			200	5.0	267	3.7										
Sept. 17, a. m....			202	5.0	188	5.3	228	4.4								
17, p. m....			168	6.0	183	5.5										
17, p. m....			302	3.3	276	3.6										
24, a. m....			153	6.5	265	3.8	294	3.4	274	3.6	270	3.7				
24, p. m....							632	1.2	357	2.8	369	2.7	289	3.5		
30, p. m....			229	4.4	209	4.8										
30, p. m....			255	3.9	250	4.0										
Oct. 5, p. m....			150	6.7	200	5.0										
5, p. m....			212	4.7												
17, p. m....			368	3.7	313	3.2	368	2.7	287	3.5						
29, a. m....			216	4.6	226	4.4	315	3.2								
29, p. m....									419	2.4						
29, p. m....			180	5.6	182	5.5	200	5.0	193	5.2						
30, a. m....			188	5.3	235	4.3										
30, p. m....			127	7.9	165	6.1										

SUMMARY.

CLEAR.

	1,000		1,500		2,000		3,000		4,000		5,000		6,000		Mean	
Morning........		8.3		5.0		4.6		4.5								5.6
Afternoon......		8.7		6.0		5.4		3.9		4.3		3.8		3.7		5.1
Mean......		8.6		5.8		5.1		4.1		4.3		3.8		3.7		5.1

CLOUDY.

Morning........		5.7		5.3		4.4		4.7		3.8		3.7				4.6
Afternoon......		7.1		5.7		5.7		4.5		3.6		2.7		3.5		4.7
Mean......		6.2		5.6		5.2		4.6		3.7		3.2		3.5		4.6

COMBINED.

Morning........		6.8		5.2		4.5		4.6		3.8		3.7				4.8
Afternoon......		8.3		5.9		5.5		4.2		4.1		3.7		3.6		5.0
Mean......		7.6		5.7		5.1		4.4		4.0		3.7		3.6		4.9

CLEVELAND, OHIO.

TEMPERATURE.

There were 93 ascensions and 227 observations at Cleveland at altitudes of 1,000 feet or over, and the highest altitude reached was 6,135 feet. The mean decrease in temperature with increase of altitude was 4.2° for each 1,000 feet, the smallest gradient in the lake region. Beginning with the highest gradient, that up to 1,000 feet, there is a quite rapid decrease up to 2,000 feet, the minimum being reached up to 3,000 feet, after which there is an increase, which becomes very slow above 4,000 feet.

But 13 observations were made in the afternoon, so that the mean results are practically those of the morning alone. The few that were made in the afternoon gave results varying from 4.9° per thousand feet up to 1,000 feet to 6° up to 2,000 feet.

The clear and cloudy weather gradients differed only slightly, but, in marked contrast to previous experiences, it was found that the cloudy-weather ones were the greater at all altitudes, with an extreme difference of 0.6° per thousand feet up to 5,000 feet.

The cloud effects were not markedly different from those at other stations. The most frequent result was a suspension of the temperature fall as the kite ascended. Once or twice there was a slight rise in the temperature, and on two occasions there was a considerable fall. On June 8 there was a fall of 4° while the kite remained in the clouds in the neighborhood of 5,000 feet elevation from 8 a. m. to 12 m. On July 28 there was a fall of 7.5° between the 3,500 and 4,400 foot levels while the kite was in the midst of the stratus clouds, and an almost similar fall, 8.5°, at the surface. The gradient, consequently, remained practically unchanged.

Twenty-one cases of inversion of temperature were found, five of them extending to the height of 2,000 feet. None were particularly marked except those of October 10 and 17. On the former date there was an inversion of 8° at the height of 1,455 feet, and on the latter date one of 10° at the height of 1,500 feet. The inversions were more marked with ESE. to SSE. winds, those from the warmest quarter, than with those from any other direction.

WINDS.

As the kite ascended its deflections were in accordance with the usual rule, with but five minor exceptions. The deflections were not marked, rarely amounting to as much as 90°, and there was a tendency toward an increase as the altitude became greater. In five cases there was a deflection in the opposite direction, amounting in one instance to 45°, and in two of them thunderstorm conditions were present.

RELATIVE HUMIDITY AND VAPOR PRESSURE.

The relative-humidity results were much the same above and at the surface, only the single observation at 6,000 feet showing any marked difference. The greatest difference, with the single exception noted, was at 5,000 feet, where the humidity was 6 per cent greater than at the surface. The approach of rain was frequently indicated by a sharp rise in the humidity as the altitude increased, but only a very few hours before the rain began. The presence of clouds almost invariably resulted in a rise in the humidity of from 5 to 10 per cent, followed by a nearly corresponding fall as the kite emerged. In a few cases the clouds were without apparent effect.

The vapor-pressure results follow below.

DIMINUTION OF VAPOR PRESSURE WITH ALTITUDE.

VALUE OF $\frac{p}{p^o}$ AT EACH RESPECTIVE 1,000 FEET OF ALTITUDE.

1,500 feet.	2,000 feet.	3,000 feet.	4,000 feet.	5,000 feet.	6,000 feet.
0.83	0.77	0.68	0.53	0.55	0.48

TEMPERATURE.

CLEAR.

Date.	1,000 feet. Rate.	Gradient.	1,500 feet. Rate.	Gradient.	2,000 feet. Rate.	Gradient.	3,000 feet. Rate.	Gradient.	4,000 feet. Rate.	Gradient.	5,000 feet. Rate.	Gradient.	6,000 feet. Rate.	Gradient.	Mean. Rate.	Gradient.
June 5, p. m....	154	6.5	219	4.6												
6, p. m.	376	2.7														
7, a. m.			782	1.3												
8, a. m.											356	2.8				
8, a. m.											172	5.8				
9, p. m.	235	4.3														
11, a. m.							590	1.7								
11, a. m.							373	2.7								
12, a. m.			216	4.6			275	3.6			252	4.0				
14, a. m.					228	4.4	236	4.2	240	4.2						
15, a. m.			135	7.4	169	5.9	239	4.2	269	3.7						
20, a. m.			443	2.3	183	5.5	353	2.8								
21, a. m.			141	7.1	154	6.5	175	5.7								
22, p. m.	166	6.0	201	5.0	166	6.0										
23, a. m.	739	1.4	203	4.9												
24, a. m.	169	5.9	494	2.0	393	2.5	357	2.8								
26, a. m.			270	3.7	341	2.9	286	3.5	254	3.9			237	4.2		
27, a. m.			474	2.1	569	1.8	421		203	4.9						
28, a. m.			182	5.5	188	5.3	215	4.7								
28, a. m.							150	6.7								
28, p. m.					165	6.1										
30, a. m.	111	9.0	630	1.6	425	2.4	313	3.2								

Temperature—Clear—Continued.

Date.	1,000 feet. Rate.	Gradient.	1,500 feet. Rate.	Gradient.	2,000 feet. Rate.	Gradient.	3,000 feet. Rate.	Gradient.	4,000 feet. Rate.	Gradient.	5,000 feet. Rate.	Gradient.	6,000 feet. Rate.	Gradient.	Mean. Rate.	Gradient.
July 1, a. m			218	4.6	246	4.1										
2, a. m			306	3.3	387	2.6	262	3.8								
2, a. m	144	6.9	143	7.0												
3, a. m			322	4.5	262	3.8	224	4.5	208	4.8						
5, a. m	135	7.4	152	6.6												
6, p. m	168	6.0														
8, a. m			171	5.8	172	5.8										
9, a. m	106	9.4														
10, a. m	139	7.2			161	6.2	147	6.8								
11, a. m			196	5.1	236	4.2										
12, p. m			182	5.5												
13, p. m	204	4.9	196	5.1												
19, a. m			258	3.9	253	4.0	186	3.4								
20, a. m			279	3.6	289	3.5	266	3.7	241	4.1	228	4.2				
21, a. m			155	6.5	290	3.4										
22, a. m	154	6.5	286	3.5												
23, a. m			310	3.2	199	5.2										
25, a. m			310	3.2	245	3.5	226	4.4								
28, a. m					409	2.4	212	4.7	211	4.7						
29, a. m			1,550	0.6	506	2.0										
31, a. m	200	5.0	178	5.6												
Aug. 3, a. m			431	2.3	403	2.5	327	3.1	256	3.9						
5, a. m	96	10.2														
7, a. m			620	1.6	466	2.1	247	4.0								
12, a. m	162	6.2	362	2.8	203	4.0										
15, a. m							432	2.3	280	3.6	312	3.2				
19, a. m	161	6.2	124	8.1												
22, a. m	316	3.2														
23, a. m							529	3.9	287	3.5						
26, a. m			156	6.4	160	6.2	211	4.7								
27, a. m			143	7.0	170	5.9										
28, a. m					447	2.2										
29, a. m							405	2.5								
31, a. m					352	2.8										
Sept. 1, a. m			142	7.0												
2, a. m			774	1.3	528	1.9										
3, a. m	464	2.2	834	3.0												
8, a. m			256	3.9	238	4.2	182	5.5								
11, a. m	119	5.0														
13, a. m							619	1.6	294	3.4						
14, a. m	267	3.7	327	3.1	305	3.3										
16, a. m			209	4.8	197	5.1										
18, a. m			288	3.5	284	3.5	357	3.9	234	4.3	222	4.5	231	4.3		
19, a. m					799	1.3	606	1.7	460	2.2	316	3.2				
25, a. m			1,012	1.0												
29, a. m	184	5.4														
Oct. 3, a. m					842	1.2	302	3.3								
6, a. m			768	1.3	280	3.6										
10, a. m							204	4.9								
12, a. m			204	3.8	277	3.6										
16, a. m	290	3.4														
19, a. m					1,480	0.7	464	2.2	240	4.2						
20, a. m							600	1.7								

CLOUDY.

Date.	1,000 feet. Rate.	Gradient.	1,500 feet. Rate.	Gradient.	2,000 feet. Rate.	Gradient.	3,000 feet. Rate.	Gradient.	4,000 feet. Rate.	Gradient.	5,000 feet. Rate.	Gradient.	6,000 feet. Rate.	Gradient.	Mean. Rate.	Gradient.
June 10, a. m			1,021	1.0			280	3.6								
19, a. m	242	4.1					257	3.9	182	5.5						
25, p. m	244	4.1														
July 4, a. m	133	7.5														
18, a. m			180	5.6	225	4.4										
19, a. m									174	5.7	185	5.4				
Aug. 3, a. m											219	4.6				
4, a. m			149	6.7	178	5.6										
16, a. m							1,455	0.7	279	3.6						
17, a. m			205	4.9	337	3.0	250	4.0								
24, a. m							477	2.1	353	2.8	271	3.7				
25, a. m			844	1.2	804	1.2	311	3.2								
26, a. m									193	5.2						
29, a. m									288	3.5						
Sept. 5, a. m			234	4.3	248	4.0	407	2.5								
6, a. m	119	8.4	130	7.7												
7, a. m			185	5.4	208	4.8	208	4.8								
10, a. m			183	5.5	187	5.3	206	4.9								
15, a. m			291	3.4	334	3.0	216	4.6								
23, a. m			270	3.7	286	3.5	286	3.5	267	3.7						
26, a. m	194	5.2														
Oct. 1, a. m			501	2.0												
5, a. m					255	3.9										
10, a. m			776	1.2	689	1.5										
14, a. m			226	4.4	229	4.4	231	4.3								
15, a. m			235	4.3	241	4.1	242	4.1								
23, a. m			245	4.1	250	4.0	279	3.6								

Temperature—Continued.

SUMMARY.

CLEAR.

Date.	1,000 feet.		1,500 feet.		2,000 feet.		3,000 feet.		4,000 feet.		5,000 feet.		6,000 feet.		Mean.	
	Rate.	Gradient.	Rate.	Gradient.	Rate.	Gradient.	Rate.	Gradient.	Rate.	Gradient.	Rate.	Gradient.	Rate.	Gradient.	Rate.	Gradient.
Morning......		5.8		4.1		3.4		3.4		4.0		4.0		4.3		4.1
Afternoon		5.1		5.0		6.0										5.4
Mean		5.6		4.1		3.5		3.4		4.0		4.0		4.3		4.1

CLOUDY.

Morning......		6.3		4.1		3.8		3.6		4.3		4.6				4.4
Afternoon		4.1														4.1
Mean		5.9		4.1		3.8		3.6		4.3		4.6				4.4

COMBINED.

Morning......		5.9		4.1		3.5		3.5		4.1		4.1		4.3		4.2
Afternoon		4.9		5.0		6.0										5.3
Mean		5.7		4.1		3.6		3.5		4.1		4.1		4.3		4.2

INVERSIONS.

Date.	1,000 feet.		1,200 feet.		1,500 feet.		2,000 feet.	
	Rate.	Gradient.	Rate.	Gradient.	Rate.	Gradient.	Rate.	Gradient.
June 10	352	2.8			1,605	0.6		
11					1,530	0.7		
Aug. 13	961	1.0						
16					539	1.9	983	1.0
24					3,180	0.3		
31					1,456	0.7		
Sept. 1	526	1.9						
9	204	4.9			411	2.4	585	1.7
13			255	3.9	832	3.0		
21	1,065	0.9						
25	463	2.2						
28			265	3.8				
30			276	3.6	408	2.5		
30								
Oct. 1	284	4.3						
1	237	4.2						
2					1,528	0.7		
3					2,910	0.3		
3					182	5.5	493	2.0
10			136	7.4	150	6.7	800	1.2
17					629	1.6		
19					602	1.7	1,979	0.5
20								

DULUTH, MINN.

TEMPERATURE.

At Duluth 96 ascensions and 328 observations were investigated, and the highest altitude attained was 6,832 feet.

The mean decrease in temperature with increase of altitude was 4.5° per thousand feet. The largest gradient, 5.2° per thousand feet, occurred up to 1,000 feet, and thereafter there was a slow and regular decrease to 3.8° up to 5,000 feet, and a more rapid increase to 4.6° up to 6,000 feet.

The morning average was slightly less, and the afternoon one slightly greater than the mean, but in neither was the rate of change regular up to the different altitudes.

The gradients in clear and cloudy weather only differed 0.1° per thousand feet, and the greatest difference in any altitude was 0.6° up to 3,000 feet.

An exceptionally large gradient was found on October 29, when, in cloudy weather, at 1 o'clock in the afternoon, the gradient was 14.9° at 1,000 feet elevation. Brisk NNW. winds were blowing from a high area to the northwestward, but no marked fall in temperature had occurred in that region, nor did any occur at Duluth.

While the cloud effects were more uniform than at most other stations, they were not so decided except in the one important instance described below. The usual result was a suspension of the temperature fall as the kite rose; at times a rise of 2° or 3°, and a decrease in the rate of fall as the kite rose above the clouds. On June 17 the temperature rose 3° as the kite reached the clouds; then fell 5° while in the clouds, and afterwards rose 1° in reaching an altitude of 4,650 feet, 1,000 feet above the top of the clouds.

A very remarkable inversion, due to clouds, occurred during the afternoon of June 10. There were cumulus clouds at an altitude of 1,391 feet, and the temperature rose steadily from 43° at the surface to 54° at the clouds, resulting in an inversion of 11°, or at the rate of 7.9° per thousand feet. The temperature remained at the highest point for about ten minutes, the elevation continuing practically the same. As the kite left the clouds in descent, the temperature fell almost instantly from 54° to 39°, the altitude being 1,038 feet.

Twenty-seven cases of temperature inversion were found, and many of them are worthy of note, not for the extent of the inversion, but for the time of day at which they occurred. Nearly one-half the cases occurred between the late morning and the middle of the afternoon, and during cloudy weather. It appears that they are almost entirely due to the effect of the easterly winds, mostly northeasterly, from Lake Superior, the warming effect of these winds being sensible at times to the height of nearly 6,000 feet. Thus, on May 26, at 1.30 p. m., there was an inversion of 3° at 5,100 feet elevation, or at the rate of 0.6° per thousand feet; on June 3, one of 2° at 5,372 feet elevation, or at the rate of 0.4° per thousand feet, and on September 20, at 4 p. m., one of 1° at 5,714 feet elevation, or at the rate of 0.2° per thousand feet. In all these cases the upper wind directions were almost diametrically opposite to those at the surface. On August 31 there was an inversion of 10° up to 4,000 feet elevation, or at the rate of 2.5° per thousand feet, and again the directions of the upper and lower currents differed 180°. On September 29 there was an inversion of 13° at 2,900 feet altitude, or at the rate of 4.5° per thousand feet; but in this case the upper winds were from the ESE. and SSE. The inversion of 15° on June 10, due to clouds, has already been noticed.

WINDS.

As a rule the wind directions differed less than usual as the kite ascended, the deflection of the latter, of course, being toward the right in a great majority of cases. The few wide divergences were toward the left, and were due to the abnormal northeast wind from the lake. At such times there would usually be an inversion of temperature, and frequently a thunderstorm a few hours later.

This northeast wind was very often purely local, being due to lake influence, and corresponding in a minor way to the sea breezes at the ocean shore. They were sometimes not more than 700 or 800 feet in depth, and rarely more than 2,000 feet. Some peculiar temperature effects, due to this east wind, were noted on July 19. The winds, both above and at the surface, were from south to southwest during the greater portion of the day, but in the afternoon an east wind set in from the lake, the upper direction remaining unchanged. This east wind was about 2,000 feet in depth, as indicated by the thermograph trace sheet, the temperature beginning to fall as soon as the kite came down into it from above, falling from 77° to 70°. The wind from the east continued until about 7 p. m., when it again changed to westerly, and the temperature rose from 67° to 80°.

RELATIVE HUMIDITY AND VAPOR PRESSURE.

The relative humidities decreased steadily, though slowly, with increase of altitude; the highest, 72 per cent, being found at 1,500 feet, and the lowest, 57 per cent, at 6,000 feet. There was very little difference between the upper air and the surface humidities, the maximum, 5 per cent, occurring at 3,000 feet, and the minimum, 1 per cent, at 6,000 feet. The surface humidities were less than those above up to 4,000 feet and greater at the higher altitudes.

Owing to the geographical situation of Duluth, at the western end of Lake Superior, the highest humidities occurred with easterly winds and the lowest with those from west to northwest.

The cloud effects upon the humidity were normal, consisting of a rise, more or less decided, but none particularly so, and a return to the preexisting conditions when the kite was freed from cloud influence. The vapor pressure results, as given below, will be found to agree almost exactly with those at Dubuque.

DIMINUTION OF VAPOR PRESSURE WITH ALTITUDE.

VALUE OF $\frac{p}{p5}$ AT EACH RESPECTIVE 1,000 FEET OF ALTITUDE.

1,500 feet.	2,000 feet.	3,000 feet.	4,000 feet.	5,000 feet.	6,000 feet.
0.82	0.79	0.74	0.64	0.57	0.45

TEMPERATURE.
CLEAR.

Date.	1,000 feet. Rate.	Gradient.	1,500 feet. Rate.	Gradient.	2,000 feet. Rate.	Gradient.	3,000 feet. Rate.	Gradient.	4,000 feet. Rate.	Gradient.	5,000 feet. Rate.	Gradient.	6,000 feet. Rate.	Gradient.	Mean. Rate.	Gradient.
May 15, p.m			318	3.1	159	6.3										
16, a.m	96	10.4	91	11.0												
23, a.m							840	1.2								
30, a.m	385	2.6														
30, p.m					120	7.8	216	4.6								
June 8, a.m	90	10.1			174	5.7										
8, p.m			113	8.8	166	6.0										
14, a.m	177	5.6	153	6.5	153	6.5	171	5.8	216	4.6	241	4.1	207	4.8		
19, a.m			536	1.9	295	3.4										
19, a.m					199	5.0										
21, a.m					042	1.1										
21, p.m					517	1.9										
28, p.m	459	2.2			309	3.2	234	4.3	282	3.5						
July 6, a.m			213	4.7	196	5.1	236	4.2	239	3.3						
6, p.m					255	3.9										
8, a.m	242	4.1	280	3.6	298	3.4	244	4.1	426	2.3						
8, p.m					94	10.7										
11, p.m			151	0.6	148	6.8	152	6.6								
12, a.m			647	1.5	380	2.6	212	4.7								
13, p.m	381	2.6	393	2.5	263	3.8	243	4.1								
14, a.m			280	3.6	228	4.4										
17, p.m	246	4.1	215	4.7	213	4.7	256	3.9	244	4.1	406	2.5				
18, a.m			201	5.0	296	3.4										
19, p.m			294	3.4	269	3.7										
20, a.m	111	9.0	106	9.4												
21, a.m			653	1.5	506	2.0										
21, a.m					160	6.0										
21, p.m			168	6.0					292	3.4	554	1.8				
22, p.m			112	8.9	177	5.6	157	6.4								
24, p.m			112	8.9	154	6.5	170	5.9	168	6.0						
25, a.m	371	2.7														
27, a.m			185	5.4	565	1.8	322	3.1	208	4.8						
29, a.m	242	4.1	110	9.1												
29, p.m	85	11.8														
Aug. 4, a.m	564	1.8														
7, a.m	552	1.8	495	2.0	558	1.8	415	2.4	250	3.9	285	3.5				
7, p.m							152	6.6								
8, a.m			241	4.1	227	4.4	188	5.3	199	5.0						
9, a.m			774	1.3	421	2.4	201	5.0								
11, a.m	108	9.3	134	7.5	144	6.9	166	6.0								
11, p.m					125	8.0	143	7.5	177	5.6						
12, a.m	333	3.0	218	4.6	142	7.0										
15, a.m							347	2.9	293	3.4	242	4.1				
16, p.m							144	6.9	159	6.3						
19, a.m			242	4.1	232	4.3	233	4.3	204	4.9						
19, p.m							162	6.2								
20, a.m			126	7.9	156	6.4										
24, p.m	173	5.8	416	2.4	480	2.1	331	3.0								
26, p.m	186	5.4														
Sept. 2, p.m	141	7.1	142	7.0	149	6.7	176	5.7	418	2.4						
3, p.m							163	6.1								
4, a.m	179	5.6	207	4.8	220	4.5	147	6.8								
4, p.m					145	6.9	160	5.9								
5, a.m					447	2.2	219	4.6								
11, a.m					315	3.2										
11, p.m							161	6.2								
16, a.m							419	2.4	255	3.9	263	3.8				
17, a.m			774	1.3	678	1.5	284	3.8	232	4.3	205	4.9				

Temperature—Clear—Continued.

Date.	1,000 feet.		1,500 feet.		2,000 feet.		3,000 feet.		4,000 feet.		5,000 feet.		6,000 feet.		Mean.	
	Rate.	Gradient.	Rate.	Gradient.	Rate.	Gradient.	Rate.	Gradient.	Rate.	Gradient.	Rate.	Gradient.	Rate.	Gradient.	Rate.	Gradient.
Sept. 17. p. m													209	4.8		
18, a. m					509	2.0	228	4.4								
18, p. m							155	6.5								
19, a. m					532	1.9										
19, a. m					317	3.2	293	3.4								
20, p. m							876	1.1								
21, a. m	1,164	0.9														
25, a. m					563	1.8	185	5.4								
27, p. m	167	6.0	147	6.8	193	5.2	182	5.0								
28, a. m	144	6.8	171	5.8	187	5.3			344	2.9						
28, p. m							521	1.9								
30, a. m	426	2.3	578	1.7	648	1.5	267	3.7								
30, p. m									238	4.2	240	4.2				
Oct. 3, a. m	170	5.9	179	5.6	171	5.8	318	3.1	371	2.7						
3, p. m											333	4.3				
6, a. m	157	6.4	202	5.0	281	3.6	254	3.9	237	4.2						
6, p. m											269	3.7	273	3.7		
7, p. m			87	11.5	96	10.4	131	7.6	153	6.5						
8, a. m					433	2.3	133	7.5	142	7.0						
10, p. m	175	5.7	180	5.6	194	5.2	229	4.4								
11, p. m	123	8.1	151	6.6	188	5.3	216	4.6	276	3.6						
14, p. m	135	7.4														
22, a. m	1,019	1.0	292	3.4	224	4.5										
23, a. m			268	3.7	256	3.9	204	4.9	196	5.1						
26, a. m	337	3.0	251	4.0	243	4.1										
31, p. m	119	8.4	157	6.4	164	6.1	277	3.6								

CLOUDY.

Date.	1,000 feet.		1,500 feet.		2,000 feet.		3,000 feet.		4,000 feet.		5,000 feet.		6,000 feet.		Mean.	
May 19. p. m			202	5.0	206	4.9										
23, p. m					167	6.0	210	4.8								
24, a. m			317	3.2	242	4.1	348	2.9								
24, p. m							848	2.9	334	3.0						
27, a. m			340	2.9	333	3.0										
June 1, a. m					332	3.0	301	3.9	313	3.2	267	3.7	242	4.1		
1, a. m									137	7.3	171	5.8	185	5.4		
1, p. m					159	6.3	158	6.3								
2, a. m			1,420	0.7												
4, p. m	390	2.6	676	1.5												
9, a. m			201	5.0	205	4.9										
9, p. m	204	4.9														
10, p. m	346	2.9														
17, a. m			306	3.3	321	3.1	486	2.1	267	3.7						
17, p. m					216	4.6	188	5.3								
21, a. m	292	3.4														
23, a. m	1,064	0.9	427	2.3	322	3.1										
24, a. m	181	5.5	205	4.9												
24, a. m	629	1.6														
July 7, a. m	343	2.9	467	2.1												
23, a. m	629	1.6														
28, a. m			129	7.8	148	6.8										
30, a. m	890	1.1														
Aug. 5, a. m			163	1.6												
16, a. m	168	6.0	200	5.0	214	4.7	156	6.4								
23, p. m	145	6.9			126	7.9										
24, a. m	252	4.0	264	3.8	296	3.4	493	2.0	352	2.8	297	3.4				
31, a. m	157	6.4														
Sept. 3, a. m	140	7.1	128	7.8	152	6.6										
5, a. m	249	4.0	536	1.9	485	2.1										
6, a. m					207	4.8	201	5.0	236	4.2						
7, p. m	164	6.1	172	5.8	183	5.5	177	5.6								
9, a. m	297	3.4	222	4.5	198	5.1										
15, p. m	320	3.1	172	5.8	188	5.3										
Oct. 1, a. m	94	10.6														
2, p. m	219	4.6	199	5.0	208	4.8	208	4.8	201	5.0						
4, a. m	243	4.1														
5, a. m	107	9.3	155	6.5												
9, a. m	170	5.9	177	5.6												
9, p. m					312	3.2										
15, a. m	353	2.8	265	3.8	258	3.9	780	1.3								
21, p. m	156	6.4	194	5.2												
24, p. m	196	5.1	176	5.7	201	5.0										
29, a. m			90	11.1												
29, p. m	67	14.9														
30, a. m	142	7.0	159	6.3	182	5.5	204	4.9								

Temperature—Continued.

SUMMARY.

CLEAR.

Date.	1,000 feet.		1,500 feet.		2,000 feet.		3,000 feet.		4,000 feet.		5,000 feet.		6,000 feet.		Mean.	
	Rate.	Gradient.	Rate.	Gradient.	Rate.	Gradient.	Rate.	Gradient.	Rate.	Gradient.	Rate.	Gradient.	Rate.	Gradient.	Rate.	Gradient.
Morning		4.8		4.1		3.6		4.3		4.3		4.1		4.8		4.3
Afternoon		6.2		6.4		6.0		5.2		4.3		3.3		4.2		5.1
Mean		5.3		5.0		4.5		4.8		4.3		3.7		4.4		4.6

CLOUDY.

Date.	1,000 feet.		1,500 feet.		2,000 feet.		3,000 feet.		4,000 feet.		5,000 feet.		6,000 feet.		Mean.	
Morning		4.6		4.5		4.2		3.5		4.2		4.3		4.8		4.3
Afternoon		5.8		4.9		5.3		5.0		4.0						5.0
Mean		5.0		4.6		4.7		4.1		4.2		4.3		4.8		4.5

COMBINED.

Date.	1,000 feet.		1,500 feet.		2,000 feet.		3,000 feet.		4,000 feet.		5,000 feet.		6,000 feet.		Mean.	
Morning		4.7		4.3		3.8		4.1		4.3		4.2		4.8		4.3
Afternoon		6.0		5.9		5.8		5.1		4.2		3.3		4.2		4.9
Mean		5.2		4.8		4.6		4.6		4.3		3.8		4.6		4.5

INVERSIONS.

Date.	800 feet.		1,000 feet.		1,200 feet.		1,500 feet.		2,000 feet.		2,500 feet.		3,000 feet.		3,500 feet.		4,000 feet.		4,500 feet.		5,000 feet.		5,700 feet.	
	Rate.	Gradient.	Rate.	Gradient.	Rate.	Gradient.	Rate.	Gradient.	Rate.	Gradient.	Rate.	Gradient.	Rate.	Gradient.	Rate.	Gradient.	Rate.	Gradient.	Rate.	Gradient.	Rate.	Gradient.	Rate.	Gradient.
May 17, a.m.			252	4.0																				
25, p.m.	770	1.3	524	1.9	568	1.8																		
25, p.m.			141	7.1																				
26, p.m.											219	4.0					392	2.6	773	1.3	1,702	0.6		
June 2, a.m.									1,991	0.5														
3, a.m.									345	2.9	482	2.1	549	1.8										
3, p.m.																	1,911	0.5			2,686	0.4		
10, p.m.					126	7.9																		
12, p.m.			540	1.9	581	1.7																		
15, p.m.					275	3.6																		
19, a.m.			266	3.8	641	1.6																		
21, a.m.									1,012	1.0														
21, p.m.													3,274	0.3										
July 16, a.m.	551	1.8																						
Aug. 15, a.m.							551	1.8	796	1.3														
30, p.m.							2,280	0.4	405	2.5	332	3.0												
31, p.m.													257	3.9			404	2.5						
Sept. 5, a.m.					2,328	0.4																		
11, a.m.			238	4.2					708	1.4														
16, a.m.					419	2.4			335	0.7														
17, a.m.					1,214	0.8																		
18, a.m.			204	4.9	654	1.5																		
19, a.m.					462	2.2																		
20, a.m.			178	5.6	256	3.9					477	2.1	1,386	0.7										
20, p.m.																							5,714	0.2
24, a.m.							258	3.9	458	2.2	1,151	0.9												
25, a.m.							476	2.1																
26, a.m.							316	3.2	607	0.6														
29, a.m.									1,529	0.7	262	0.8												
29, p.m.							201	5.0			207	4.8	269	3.7	219	4.6								
Oct. 8, a.m.							474	2.1																
15, a.m.																	912	1.1	790	1.3				

LANSING, MICH.

TEMPERATURE.

At Lansing 58 ascensions and 131 observations at altitudes above 1,000 feet were considered, and the greatest altitude attained was 5,415 feet. Observations were first taken on June 7. Previous to this date, however, almost daily ascensions were made without the meteorograph and observations of wind direction taken. During the ascension of May 1 an altitude of 7,068 feet was attained.

The mean gradient was 5° per thousand feet, the highest gradient of the lake region stations. There was a steady decrease with increase of altitude; it is quite rapid up to 3,000 feet and slow thereafter. As there were but 17 observations made in the afternoon, the morning results differ but slightly from the means. The few afternoon observations gave gradients varying from 3.9 per thousand feet up to 2,000 feet to 6.2 up to 1,500 feet.

The clear and cloudy weather gradients did not differ greatly except up to 3,000 feet, where there was a difference of 1.5° per thousand feet, the clear-weather gradient—4.5°—being the greater. The clear-weather gradients were uniformly the greater, except up to 4,000 feet. When clouds enveloped or were near the kite the usual temperature rises where noticed, except on one day—August 25—when there was a fall of 3.7° while the kite was descending in dense cumuli from about 2,100 feet altitude to 1,900 feet. On September 26 there was a rise of 8° while the kite was rising in cumulus clouds from 2,700 to 3,200 feet, resulting in a complete inversion of the gradient. There was also an inversion of 1° at an elevation of 2,300 feet on October 12, while the kite was in dense cumuli just previous to the commencement of rain. The direction of the wind appeared to exercise some effect upon the gradients. The greatest were found with the warm winds from the southwest, and the least with the colder ones from the northwest. There were nine days on which inversions of temperature occurred, and on four of them the results were quite marked. On September 29 there was an inversion of 10° at the height of 1,623 feet at 7.38 a. m., steadily decreasing after 7.30 a. m., but was still at 1.5° at 9.54 a m. at an altitude of 1,052 feet. The winds were from the south at the surface and from the southwest above. On October 1 the inversion amounted to 7° at the height of 3,200 feet at 7.50 a. m. and to 5° at 3,500 feet with winds from the east at the surface and from the east-southeast above. On October 29, with east-southeast winds at the surface and southwest to south above, there was an inversion of 13.3° at 8.18 a. m. at the height of 1,082 feet, and at 10.02 a. m. there was still an inversion of 1°.

WINDS.

Wind directions conformed to the usual rule of deflection toward the right in the upper air, but not at all to a marked degree. There were about one dozen exceptions to the rule, but none were decided, neither were they preceded, attended, nor followed by any characteristic phenomena.

RELATIVE HUMIDITY AND VAPOR PRESSURE.

The relative humidity results did not differ greatly. Up to 3,000 feet they were lower than at the surface; the greatest deficiency, 9 per cent, being found at 1,500 feet, and the least, 3 per cent, at 3,000 feet. At 4,000 feet the humidity was 7 per cent greater than at the surface, while at 5,000 feet there was a difference of but 1 per cent. Clouds as a rule caused a slight rise in the humidity, but in at least two instances there was a marked fall while the kite was enveloped in clouds.

The vapor-pressure results follow below. It will be noticed that there is a very satisfactory agreement with the figures for the lake stations at Cleveland, Sault Ste. Marie, and Duluth, at least below the 4,000-foot level.

DIMINUTION OF VAPOR PRESSURE WITH ALTITUDE.

VALUE OF $\frac{p}{p_0}$ AT EACH RESPECTIVE 1,000 FEET OF ALTITUDE.

1,500 feet.	2,000 feet.	3,000 feet.	4,000 feet.	5,000 feet.
0.87	0.83	0.73	0.56	0.51

TEMPERATURE.

CLEAR.

Date.	1,000 feet.		1,500 feet.		2,000 feet.		3,000 feet.		4,000 feet.		5,000 feet.		Mean.	
	Rate.	Gradient.	Rate.	Gradient.	Rate.	Gradient.	Rate.	Gradient.	Rate.	Gradient.	Rate.	Gradient.	Rate.	Gradient.
June 8, p.m.			127	7.9										
9, a.m.											233	4.3		
9, a.m.											281	3.6		
14, a.m.							172	5.8						
15, a.m.							190	5.3						
15, a.m.							158	6.3						
15, a.m.	103	9.7	131	7.6			214	4.7	210	4.8				
18, a.m.	150	6.7					143	7.0	167	6.0			214	4.7
19, a.m.	128	7.8			148	6.8								
21, a.m.	81	12.3	200	5.0										
23, a.m.							428	2.3	211	4.7	212	4.7		
24, a.m.							142	7.0						
24, p.m.											717	1.4		
25, a.m.							155	6.5						
26, a.m.							162	6.2						
26, a.m.			189	5.3										
28, a.m.	62	16.1	131	7.6	162	6.2								
28, a.m.			126	7.9			209	4.8	314	3.2				
30, a.m.			149	6.7	146	6.8	166	6.0						
July 1, a.m.														
1, p.m.	148	6.8												
2, a.m.	96	10.4	133	7.5	614	1.6	276	3.6	257	3.9				
3, a.m.									218	4.1				
4, a.m.	159	7.2	121	8.3	161	6.2								
4, a.m.	94	10.6												
9, a.m.							206	4.9	205	4.9				
10, a.m.	121	8.3			176	5.7	181	5.5						
18, a.m.			147	6.8	191	5.2	200	5.0						
20, a.m.			177	5.6	159	6.3	200	5.0	212	4.7				
29, a.m.	119	8.4												
Aug. 7, p.m.	201	4.9	151	6.6										
11, a.m.							180	5.6						
21, p.m.			127	7.9										
29, a.m.	137	7.3	450	2.2	187	5.3								
28, a.m.							174	5.7						
31, a.m.							417	2.4						
Sept. 2, a.m.									217	4.6				
3, a.m.									233	4.3	220	4.5		
4, a.m.									280	3.6	234	4.3		
7, a.m.							372	2.7						
8, a.m.							698	1.4						
18, a.m.	326	3.1			926	1.1	1,166	0.9	318	3.1				
19, a.m.									357	2.8	290	3.4		
26, a.m.									616	1.6				
28, a.m.					174	5.7								
30, a.m.	611	1.6							727	1.4	306	3.3		
Oct. 10, a.m.			194	5.2	273	3.7	353	2.8						
15, a.m.	159	6.3												
28, a.m.	357	2.8			333	3.0	316	3.2						
28, p.m.			337	3.0	400	2.5	434	2.3						
29, a.m.			336	3.0										
29, p.m.	174	5.7			189	5.3	208	4.8	239	4.2				

CLOUDY.

Date.	1,000 feet.		1,500 feet.		2,000 feet.		3,000 feet.		4,000 feet.		5,000 feet.		Mean.	
	Rate.	Gradient.	Rate.	Gradient.	Rate.	Gradient.	Rate.	Gradient.	Rate.	Gradient.	Rate.	Gradient.	Rate.	Gradient.
June 12, a.m.			229	4.4					270	3.7	277	3.6		
12, a.m.									203	4.1				
13, a.m.	95	10.5	131	7.6										
20, p.m.							148	6.8	157	6.4				
29, a.m.					258	3.9	431	2.3	280	3.6				
July 19, a.m.			394	2.5										
25, p.m.	171	5.8	136	7.4										
28, a.m.	143	7.0			239	4.2								
28, a.m.	121	8.3	134	7.5	185	5.4								
Aug. 16, a.m.							272	3.7						
Sept. 13, a.m.							775	1.2	237	4.2				
Oct. 1, a.m.	250	4.0					1,460	0.7						
2, a.m.														
3, a.m.					363	2.8								
12, a.m.			180	5.6	160	6.2	272	3.7						
13, a.m.					196	5.1	276	3.6	342	2.9				
23, a.m.					184	5.4								
23, a.m.							500	2.0						
23, a.m.							339	2.9						

Temperature—Continued.

SUMMARY.

CLEAR.

Date.	1,000 feet.		1,500 feet.		2,000 feet.		3,000 feet.		4,000 feet.		5,000 feet.		Mean.	
	Rate.	Gradi-ent.	Rate.	Gradi-ent.	Rate.	Gradi-ent.	Rate.	Gradi-ent.	Rate.	Gradi-ent.	Rate.	Gradi-ent.	Rate.	Grndi-ent.
Morning		7.9		6.2		4.9		4.4		3.8		3.8		5.2
Afternoon		5.8		5.8		3.9		5.0		4.2				4.9
Mean		7.6		6.1		4.8		4.5		3.9		3.8		5.1

CLOUDY.

Morning		7.4		5.5		4.7		2.5		3.7		3.6		4.6
Afternoon		5.8		7.4				6.8		6.4				6.6
Mean		7.1		5.8		4.7		3.0		4.2		3.6		4.7

COMBINED.

Morning		7.8		6.0		4.8		3.9		3.8		3.8		5.0
Afternoon		5.8		6.2		3.9		5.3		5.3				5.3
Mean		7.5		6.0		4.7		4.1		3.9		3.8		5.0

SAULT STE. MARIE, MICH.

TEMPERATURE.

At Sault Ste. Marie there were in all 74 ascensions and 180 observations at altitudes of 1,000 feet or more, and the greatest altitude attained was 5,607 feet.

The mean decrease of temperature with increase of altitude was 4.9° for each 1,000 feet. The greatest rate was found, as usual, up to 1,000 feet, while the lowest occurred up to 5,000 feet, there being a steady, though by no means regular, decrease with increase of altitude.

The morning gradients were markedly greater than those of the afternoon, except up to 5,000 feet, where it was only 0.4° less. The greatest difference occurred up to 1,000 feet, where it amounted to 4° per 1,000 feet. The afternoon results were greater than the adiabatic rate at all elevations up to 3,000 feet, becoming rapidly smaller beyond. Up to 1,000 feet the gradient was 10°, while up to 5,000 feet it was but 3.4° per 1,000 feet. It should be noted, however, that there were but seven observations at 1,000 feet elevation.

The gradients in clear weather were greater than the cloudy weather ones except up to 5,000 feet, where they were almost exactly alike. The greatest difference again occurs up to 1,000 feet, where it was 3.3° per 1,000 feet.

The presence of clouds as a rule produced the usual effects of suspension of the temperature fall with increase of elevation, at times changing to a slight rise. There would be alternate rises and falls as the kite moved in and out of the clouds, but between very limited extremes, the elevation evidently being without effect as long as clouds were present.

Inversions were very infrequent and of no consequence.

WINDS.

The kite deflections toward the right were frequently quite decided, amounting at one time to 160°, and there were many which equaled or exceeded 90°. The deflections toward the left were only three or four in number and the amount was small. Two of these abnormal deflections were followed by rain within a short time.

RELATIVE HUMIDITY AND VAPOR PRESSURE.

The relative humidity changed but little above the height of 1,500 feet, the extreme range of the mean values from 1,500 to 5,000 feet having been but 5 per cent, the minimum occurring at 4,000 feet and the maximum at 5,000 feet. The humidity at the surface decreased steadily as each day's ascension progressed, and it was higher than those at the different elevations up to 3,000 feet, the greatest difference being 9 per cent at 1,500 feet. Above 3,000 feet the humidity was higher than at the ground, with a difference of 14 per cent at 5,000 feet. Cloud effects were usually well-defined, consisting in a marked rise in the humidity as the kite entered the clouds, generally to above 90 and often to 100 per cent, followed by an equally marked fall as it emerged, the fall sometimes being as much as 35 per cent.

Vapor pressure results are given below. They agree very closely with those at Duluth and Dubuque up to 3,000 feet, but differ from 0.05 to 0.13 above that height.

DIMINUTION OF VAPOR PRESSURE WITH ALTITUDE.

VALUE OF $\frac{p}{p^0}$ AT EACH RESPECTIVE 1,000 FEET OF ALTITUDE.

1,500 feet.	2,000 feet.	3,000 feet.	4,000 feet.	5,000 feet.
0.83	0.76	0.71	0.71	0.44

TEMPERATURE.

CLEAR.

Date.	1,000 feet.		1,500 feet.		2,000 feet.		3,000 feet.		4,000 feet.		5,000 feet.		Mean.	
	Rate.	Gradient.	Rate.	Gradient.	Rate.	Gradient.	Rate.	Gradient.	Rate.	Gradient.	Rate.	Gradient.	Rate.	Gradient.
May 4, a. m			483	2.1					285	3.5				
5, a. m			144	6.9										
5, p. m			218	4.6			173	5.8	211	4.7				
5, p. m									120	8.3				
6, p. m			180	5.6	193	5.2	268	3.7						
7, a. m					448	2.2	275	3.6						
8, a. m			343	2.9										
9, p. m			128	7.8					281	3.6				
9, p. m									208	4.8				
10, a. m					154	6.5								
13, a. m			180	5.6										
15, p. m			142	7.0			153	6.5	184	5.4	275	3.6		
16, a. m					385	2.6			481	2.1	539	1.9		
17, a. m					213	4.1	285	3.5						
17, p. m							139	7.2						
24, a. m			197	5.1			222	4.5	268	3.7	274	3.6		
25, p. m			274	3.6										
26, p. m			90	11.1										
30, a. m					203	4.9								
30, a. m					386	2.6								
31, a. m			393	2.5										
June 4, p. m					154	6.5								
8, a. m					194	5.2	234	4.3						
July 15, a. m	122	8.2												
20, a. m					245	4.1			287	3.7				
21, a. m					193	5.1								
23, p. m			70	12.7										
23, p. m			72	13.9	95	10.5								
25, a. m			124	7.8										
27, a. m			92	10.9	117	8.5	165	6.1	172	5.8				
29, a. m							365	2.7						
29, a. m			202	5.0	196	5.1	261	3.8						
31, p. m			179	5.6										
Aug. 4, p. m							175	5.7						
4, p. m					116	8.6	146	6.8						
5, p. m			86	11.8	101	9.9	124	8.1	156	6.4				
8, a. m					168	6.0	179	5.6	245	4.1				
12, a. m					187	4.3	171	5.8						
12, p. m									169	5.9				
26, p. m					171	5.8	136	7.4						
30, a. m							192	5.2	239	2.9				
31, a. m							1,271	0.8	670	1.5				
31, a. m							555	1.8	881	1.1	738	1.4		
31, p. m									790	1.3				
Sept. 1, a. m							150	6.3						
3, a. m			123	8.1			175	5.7	190	5.3	203	4.9		
3, p. m									178	5.6				
11, a. m			286	3.5	346	2.9								
12, a. m					443	2.3	253	4.0						
13, a. m	117	8.5	160	6.2	189	5.3	213	4.7	202	5.0	304	3.3		
18, a. m			172	5.8	193	5.2	252	4.0						
28, a. m			193	8.1	148	6.8	162	6.2						
28, p. m									195	5.1				
28, p. m									247	4.0				
29, a. m	115	8.7			220	4.5	328	3.0	410	2.4	439	2.3		
Oct. 1, a. m			318	3.1					1.165	0.9	491	2.0		
8, a. m					169	5.9	217	4.6	221	4.5	242	4.1		

Temperature—Continued.

CLOUDY.

Date.	1,000 feet.		1,500 feet.		2,000 feet.		3,000 feet.		4,000 feet.		5,000 feet.		Mean.	
	Rate.	Gradient.	Rate.	Gradient.	Rate.	Gradient.	Rate.	Gradient.	Rate.	Gradient.	Rate.	Gradient.	Rate.	Gradient.
May 2, p. m					1,060	0.9								
3, p. m			403	2.5										
13, p. m					500	2.0	200	3.3						
18, a. m							950	1.1						
19, p. m					169	5.9					374	2.7		
20, a. m							207	4.8	370	2.7	217	4.6		
27, a. m					173	5.8								
29, p. m			148	6.8	156	6.4								
June 1, p. m			218	4.6										
3, p. m					363	2.8	425	2.4						
3, a. m									172	5.8	408	2.5		
5, p. m			235	4.3	290	3.4								
6, p. m			176	5.7										
9, p. m	100	10.0												
July 16, a. m	235	4.3	304	3.3	202	5.0								
19, a. m					118	8.5								
Aug. 1, p. m			94	10.6	130	7.7								
13, a. m					191	5.2								
15, a. m			188	6.0										
16, p. m			82	12.2	114	8.8	121	8 3						
17, p. m			120	8.8										
21, p. m			203	4.9	241	4.1	229	4.4						
28, a. m			401	2.5										
29, a. m							195	5.1	282	3.5	327	3.1		
Sept. 9, a. m			215	4.7										
25, a. m					224	4.5	240	4.2	309	3 2				
25, a. m											380	2.6		
30, p. m			143	7.0	194	5.2								
30, p. m					132	7.6								
Oct. 3, p. m					851	1.2								
16, a. m					210	4.8	1,187	0.8	621	1.6				
16, a. m							692	1.4						
23, a. m	235	4.3	240	4.2	192	5.2								
28, a. m	478	2.1	364	2.7										
29, a. m			384	2.6	392	2.6	370	2.7	459	2.2				
29, p. m											327	3.1		

SUMMARY.

CLEAR.

Morning		8.5		6.2		4.7		4.3		3.2		2.9		5.0
Afternoon				7.4		7.8		6.4		5.0		3.6		6.6
Mean		8.5		6.7		5.4		4.9		4.1		3.0		5.4

CLOUDY.

Morning		3.6		3.7		5.2		2.9		3.2		3.1		3.6
Afternoon		10.0		6.7		4.7		4.6				3.1		5.8
Mean		5.2		5.5		4.9		3.5		3.2		3.1		4.2

COMBINED.

Morning		6.0		5.4		4.8		3.9		3.3		3.0		4.4
Afternoon		10.0		7.0		5.7		5.8		5.0		3.4		6.2
Mean		6.6		6.2		5.2		4.5		3.9		3.0		4.9

DODGE CITY, KANS.

TEMPERATURE.

One hundred and thirty-eight ascensions and 573 observations were made at Dodge City at altitudes of 1,000 feet or more, and the greatest altitude attained was 8,019 feet.

The mean temperature decrease was at the rate of 4.2° for each 1,000 feet of ascent. The average decrease from the ground up to 1,000 feet was 6.3°; for the other altitudes it was as follows: Up to 1,500, 5.2° per thousand feet; 2,000 feet, 4.8°; 3,000 feet, 3.7°; 4,000 feet, 3.1°; 5,000 feet, 3.2°; 6,000 feet, 3.2°; 7,000 feet, 3.2°; 8,000 feet, 4.9°. (Plate 3.) These are the general mean values obtained from observations made in all kinds of weather and at all hours between 8 a. m. and 8 p. m., Eastern time. They are somewhat greater than those at Washington up to 4,000 feet, and, as was the case at the latter place, there is a decrease up to this height, and thereafter a practically uniform rate, agreeing almost exactly with the Washington results, except up to 8,000 feet. The difference at 8,000 feet was no doubt largely due to the time of day at which the observations were taken, that at Washington having been taken about 8 a. m., while that at Dodge City was taken in the afternoon. Both days were clear.

The inversions of temperature, of which 52 cases were noted, occurred mostly on clear days with the lighter winds, not often less than 10 miles per hour, however, at the surface, and in two cases over 20 miles. They were at a minimum during May, and gradually increased in number during the remaining months. They quite frequently extended to the height of 3,000 feet, and on one occasion to over 5,000 feet, the inversion in this case amounting to over 11°, or 2.2° per thousand feet.

Inversions were sometimes caused by the formation of clouds in the early morning. Below the clouds there would be little or no change in the temperature, while above there would be an increase on account of the direct action of the sun's rays. The result was, of course, an inversion. A marked case occurred on October 1, when there was a rise in the temperature of 11.5° within a few minutes after the kite emerged from the upper surface of the clouds.

The amount of inversion at Dodge City was in general less than at Washington on account of the higher wind velocities at the former place.

The mean morning gradient was 3.8° per thousand feet, and the same suspension of decrease above 4,000 feet occurs. The greatest gradient was found up to 1,000 feet, where it was 6.4° per thousand feet, and the least, 2.6°, up to 4,000 feet. (Plate 1.)

The afternoon gradient was, of course, much more regular, but, except up to 1,500 feet, lacks the approach to the adiabatic line that characterized the afternoon curve at Washington. The average was 4.7° per thousand feet, 1.3° less than the Washington afternoon figures and 0.7° less than the adiabatic rate. (Plate 2.) Up to 1,500 feet the morning, afternoon, and mean values are very nearly alike, reaching their greatest divergence up to 4,000 feet, and again becoming very nearly equal up to 6,000 and 7,000 feet.

As a rule the presence of clouds produced the usual phenomena of decided decreases in the gradient, at times amounting to a complete, though slight, inversion. On May 16 there was a rise in the temperature of 2.5° while the kite was in the clouds, and a return to the previous reading as soon as the edge of the clouds was reached. On May 25, while the kite was enveloped in cloud, the temperature decrease was but 1° for 2,554 feet of altitude, or 0.4° per thousand feet, while just before reaching the clouds the decrease was 1° in 943 feet, or 1.1° per thousand feet. On May 28, while the kite was at the edge of the clouds, the decrease in temperature was at the rate of 1° in 6,248 feet, or somewhat less than 0.2° per thousand feet, the altitude being 3,124 feet. On August 4, from 8 to 10 a. m., the kite reaching an altitude of 3,700 feet, there was no change in the temperature on account of clouds passing under or over the kite; then a rise of 2.3° at 10.10 a. m. while the kite was above the cloud, followed by a fall of 2° as soon as the clouds were reached in descent, and a corresponding rise when the kite emerged below. In this case the cloud temperature exhibited remarkable uniformity, remaining at about 64° with variations of less than one-half degree. A similar instance of uniformity of cloud temperature occurred on August 17, the variation, while the kite was in or near the clouds, having been but 1°. In a few cases the presence of clouds did not appear to exercise an appreciable effect upon the temperature.

Brisk to high winds were the rule, both above and at the surface, during the entire period in which ascensions were made, and the higher ones were usually, though not uniformly, attended by a decrease in the gradient. The directions of the wind did not appear to have any bearing upon the result.

WINDS.

The wind directions were practically the same above and at the surface, the differences being confined to a deflection of the kite toward the right. Frequently the deflection increased with the altitude, but rarely amounted to more than 90°. In some instances the deflection of the kite was toward the left, but neither frequently nor decidedly, except in one instance. On October 4, the day on which the greatest altitude was attained, the upper and lower winds were about the same, southwest to southwest and south, until shortly after 3 p. m., when the surface wind changed to north, again changing to northeast shortly before 5 p. m. and remaining at that point. The upper wind continued to blow from the southwest until 5 p. m., when it began to shift toward the left, reaching the northeast direction at 5.10 p. m., and again coinciding with the surface wind. The velocity of the upper wind was comparatively light after the surface wind changed to north, and remained so until both again blew from the same direction.

RELATIVE HUMIDITY AND VAPOR PRESSURE.

Generally speaking there was comparatively little change in the relative humidity with changes of elevation except in the few observations above 7,000 feet. Between the limits of 1,500 and 4,000 feet, where more than 90 per cent of the observations were made, the mean decrease was only 5 per cent, and up to 6,000 feet but 10 per cent. The differences between the surface humidities and those at the kite were also quite small. The highest humidities were noted with winds at the surface from north to northeast, and the lowest with those from southeast to west, particularly with those from southeast to south. Approaching rain, or the fact that rain had fallen a few hours previously, appeared to have little or no effect except in a few instances, and in these an equal rise took place at the same time at the surface.

Cloud effects were also comparatively insignificant except on a few days. There would be a slight rise when the kite entered a cloud, usually of 3 or 4 per cent, occasionally of 10, and once of 17 per cent; but at no time did the humidity rise as high as 90 per cent if it had been below that point before the cloud was reached. In one or two cases the humidity fell slightly while the kite was in the clouds, while on October 1 there was a fall of 45 per cent as the kite was leaving the cloud from above. It will be remembered that October 1 was also the day on which occurred the marked temperature inversion above the cloud.

The vapor-pressure results obtained were higher than those at Washington, with the exception of those at 1,500 feet. The figures for both places are given below, and it is readily apparent from an inspection of them that the changes with increase of altitude are much less at Dodge City than at Washington.

DIMINUTION OF VAPOR PRESSURE WITH ALTITUDE.

VALUE OF $\frac{p}{p^5}$ AT EACH RESPECTIVE 1,000 FEET OF ALTITUDE.

Station.	1,500 feet.	2,000 feet.	3,000 feet.	4,000 feet.	5,000 feet.	6,000 feet.	7,000 feet.	8,000 feet.
Dodge City	0.85	0.84	0.80	0.71	0.61	0.56	0.51	0.55
Washington	0.87	0.82	0 66	0.60	0.54	0.46	0.45	0.34

TEMPERATURE.

CLEAR.

Date.	1,000 feet.		1,500 feet.		2,000 feet.		3,000 feet.		4,000 feet.		5,000 feet.		6,000 feet.		7,000 feet.		8,000 feet.		Mean.	
	Rate.	Gradient.	Rate.	Gradient.	Rate.	Gradient.	Rate.	Gradient.	Rate.	Gradient.	Rate.	Gradient.	Rate.	Gradient.	Rate.	Gradient.	Rate.	Gradient.	Rate.	Gradient.
May 2, a. m...			2,270	0.4																
6, a. m...							270	3.7												
7, p. m...					201	5.0	217	4.6												
9, p. m...					126	7.9														
10, a. m...	109	9.2			166	6.0	169	5.9												
11, a. m...	153	6.5	153	6.5	140	7.1	161	6.2	172	5.8										
12, a. m...							801	1.3												
18, p. m...	119	8.4					159	8.3												
18, p. m...							234	4.3												
19, p. m...			120	8.3			150	6.7												
21, p. m...			154	6.5	182	5.5	180	5.6												
22, p. m...	174	5.7	164	6.1																
23, a. m...							1,111	0.9	443	2.8										
24, p. m...					212	4.7			277	3.6										
27, a. m...	250	4.0																		
29, a. m...					200	5.0	250	4.0	717	1.4			573	1.7						
30, a. m...					576	1.7														
31, a. m...							230	4.3			271	3.7								

Temperature—Clear—Continued.

Date.	1,000 feet. Rate.	Gradient.	1,500 feet. Rate.	Gradient.	2,000 feet. Rate.	Gradient.	3,000 feet. Rate.	Gradient.	4,000 feet. Rate.	Gradient.	5,000 feet. Rate.	Gradient.	6,000 feet. Rate.	Gradient.	7,000 feet. Rate.	Gradient.	8,000 feet. Rate.	Gradient.	Mean. Rate.	Gradient.
June 1, p.m.			130	7.7			165	6.1												
2, a.m.					148	6.8					847	1.2	505	2.0						
2, a.m.											309	3.2								
14, p.m.	139	7.2																		
19, p.m.	112	8.9	131	7.6																
20, a.m.	116	8.6	170	5.9	193	5.2			231	4.3										
20, a.m.	147	6.8	117	8.5	132	7.6														
21, a.m.									1,169	0.9										
21, a.m.			317	3.2	446	2.3			425	2.4										
23, a.m.	137	7.3	296	3.4	441	2.3			1,207	0.8										
23, a.m.			655	1.3	674	1.5			340	2.9	364	2.7	309	3.2	263	3.8				
23, a.m.					209	4.8			206	4.9	208	4.8	217	4.6						
23, p.m.	190	5.3																		
24, a.m.							627	1.6	212	4.7										
25, a.m.							2,134	0.5	281	3.6	858	1.2								
25, a.m.					170	5.9														
26, a.m.	158	6.3	148	6.8	161	6.2														
26, a.m.			163	6.1	139	7.2														
27, p.m.	150	6.7	125	8.0																
27, a.m.			154	6.5	167	6.0														
27, a.m.			139	7.2																
28, a.m.	118	8.5			574	1.7			396	2.5										
28, a.m.			135	7.4	150	6.7			261	3.8										
29, a.m.									1,343	0.7	460	2.1	368	2.7						
29, a.m.	116	8.6	140	7.1	150	6.7			158	6.3	247	4.0								
30, a.m.									959	1.0	392	2.6								
30, a.m.			133	7.5	164	6.1			257	3.9										
July 1, a.m.	129	7.8			404	2.5														
1, a.m.			150	6.7	189	5.3														
2, a.m.	996	1.0			510	2.0	349	2.9			273	3.7								
2, a.m.			128	7.0	130	7.7	155	6.5												
3, p.m.	120	8.3	149	6.7																
5, a.m.	138	7.2	167	6.0	179	5.6	187	5.3												
5, a.m.	114	8.8			143	7.0														
6, a.m.							1,029	1.0	464	2.2										
6, a.m.	135	7.4	144	6.9			277	3.6	366	2.7										
9, a.m.	165	6.1	193	5.2	131	7.6														
9, a.m.	110	8.4	125	8.0																
11, a.m.	246	4.1	370	2.7	235	4.3														
12, p.m.	194	5.2	194	5.2	193	5.2														
12, p.m.			172	5.8																
13, p.m.	138	7.2																		
13, p.m.	146	6.8																		
15, r.m.			227	4.4	157	6.4														
16, a.m.			252	4.0	250	4.0														
17, a.m.	190	5.3	197	5.1																
17, a.m.							1,104	0.9	514	1.9	348	2.9								
17, a.m.	151	6.6	163	6.1	145	6.9			260	3.8										
19, a.m.	170	9.3	206	4.9	475	2.1	1,977	0.5												
20, a.m.			164	6.1	205	4.9														
20, a.m.			153	6.5																
20, p.m.	246	4.1																		
22, a.m.			445	2.2																
22, a.m.			325	3.1	404	2.5														
23, a.m.	514	1.9			251	4.0														
23, p.m.			458	2.2	371	2.7														
25, a.m.	199	5.0																		
26, a.m.	173	5.8	180	5.6	190	5.3	199	5.0												
26, a.m.					169	5.0	178	5.6	207	4.8										
26, a.m.					530	1.9														
28, a.m.			1,343	0.7																
28, a.m.	130	7.7	315	3.2																
29, a.m.			446	2.2			2,533	0.4												
29, a.m.			181	5.5	394	2.5	986	1.0												
30, a.m.	118	8.5			3,029	0.3	588	1.7												
30, a.m.	85	11.8	180	5.6			336	3.0												
Aug. 8, a.m.							683	1.5	287	3.5										
8, a.m.					177	5.6	215	4.7												
10, a.m.	239	4.2	289	3.5																
12, a.m.			1,313	0.8	527	1.9	450	2.2	305	3.3										
12, a.m.	122	8.2			203	4.9	252	4.0												
13, a.m.	107	9.3	186	5.4			1,057	0.9												
15, a.m.	247	4.0																		
16, a.m.					3,175	0.5														
19, a.m.	295	3.4	283	3.5	244	4.1	254	3.8												
19, p.m.	138	7.2	198	5.1																
20, a.m.	122	8.2			227	4.4	313	3.2												
21, a.m.			260	3.8	258	3.9	220	4.5												
21, p.m.	181	5.5			194	5.2														
22, a.m.	126	7.9	168	6.0	295	4.4	273	3.7	377	2.7										
22, a.m.	171	5.8							699	1.4										
23, a.m.			260	3.7	462	2.2	418	2.4	429	2.3										
24, a.m.			435	2.3	404	2.5	363	2.8												
24, a.m.	153	6.5	245	4.1																
25, p.m.			191	5.2																
26, p.m.	144	6.9																		
26, a.m.			172	5.8	189	5.3														
28, a.m.	208	4.8	363	2.8	413	2.4	937	1.1												
29, a.m.	148	6.8	215	4.7	140	7.1	241	4.1												
30, a.m.									980	1.1	343	2.9	247	4.0						

Temperature—Clear—Continued

Date.	1,000 feet.		1,500 feet.		2,000 feet.		3,000 feet.		4,000 feet.		5,000 feet.		6,000 feet.		7,000 feet.		8,000 feet.		Mean.	
	Rate.	Gradient.	Rate.	Gradient.	Rate.	Gradient.	Rate.	Gradient.	Rate.	Gradient.	Rate.	Gradient.	Rate.	Gradient.	Rate.	Gradient.	Rate.	Gradient.	Rate.	Gradient.
Aug. 30, a. m...	144	8.9	161	6.2			155	6.7	197	5.1	231	4.3								
31, a. m...	150	6.7	173	5.8	200	5.0	370	2.7												
Sept. 1, a. m...					338	3.0	478	2.1												
1, a. m...	153	6.5	208	4.8	252	4.0														
2, a. m...	161	6.2	180	5.6	751	1.3	1,427	0.7												
3, a. m...									1,147	0.9										
3, a. m...	120	8.3	156	6.4	177	5.6	299	3.3	468	2.1										
4, a. m...											639	1.5	433	2.3	330	3.0				
4, a. m...	121	8.8	115	8.7					355	3.9	284	3.5	296	3.4						
5, a. m...			260	3.8					2,153	0.5	596	1.7								
5, a. m...	106	9.4	144	6.9			305	3.3	348	2.9										
7, p. m...	380	2.6	401	2.5	741	1.3														
8, a. m...			771	1.3	708	1.4	357	2.8	311	3.2										
8, a. m...	143	7.0	143	7.0	143	7.0	192	5.2												
14, p. m...	354	2.8	306	3.3	222	4.5	260	3.8												
14, p. m...					214	4.7	243	4.1												
17, a. m...	667	1.5	1,188	0.8																
30, a. m...			258	3.9	202	5.0	226	4.4	340	4.2										
20, a. m...					164	6.1	175	5.7												
21, a. m...			528	1.9	402	2.5	265	3.8	253	4.0										
21, p. m...	108	9.3			170	5.9	161	6.2												
22, a. m...	258	3.9																		
23, a. m...	150	6.7			196	5.1	192	5.2	198	5.1	199	5.0								
23, a. m...					228	4.4														
23, p. m...					159	6.3	184	5.4												
27, p. m...			190	5.3	172	5.8	179	5.6	200	5.0										
27, p. m...	236	4.2	212	4.7	198	5.1														
28, a. m...	195	5.1	149	6.7	150	6.7	340	2.9	904	1.1										
29, p. m...	122	8.2	135	7.4	147	6.8														
29, p. m...					172	5.8														
29, p. m...			126	7.9	146	6.8														
30, a. m...			264	3.8	270	3.7														
30, a. m...					661	1.5	472	2.1	213	4.7	277	3.6	304	3.3						
30, p. m...	128	7.8	133	7.5			180	5.6												
Oct. 1, a. m...	196	7.4																		
2, a. m...	133	7.5	299	3.3																
3, a. m...	525	1.9																		
4, a. m...									2,758	0.4	297	3.4	253	4.0						
4, p. m...													228	4.4	296	3.5	204	4.9		
4, p. m...	159	6.3	360	2.8					1,635	0.6			345	2.9						
5, a. m...	167	6.0																		
5, a. m...	118	8.5	133	7.5	229	4.2	1,310	0.8			348	2.9	446	2.2						
6, p. m...			180	5.6	197	5.1	192	5.2			515	1.9								
6, p. m...	2,364	0.4			294	3.4	833	3.0	311	3.2	250	4.0								
9, a. m...									1,866	0.5										
9, p. m...			214	4.7	243	4.1	238	4.2												
10, p. m...			155	6.5	159	6.3														
10, p. m...			142	7.0																
11, p. m...	145	6.9	159	6.3	154	6.5	177	5.6	202	5.0	201	5.0								
11, p. m...					506	2.0	813	3.2	249	4.0										
12, a. m...			148	6.8	152	6.6	116	8.6	304	3.3	1,624	0.6								
14, a. m...	118	8.5	122	8.2	146	6.8	284	3.8												
15, a. m...									2,310	0.4	658	1.5								
15, p. m...	109	9.2	148	6.8	151	6.6	184	5.4	315	3.2	358	2.8								
18, a. m...							1,034	1.0												
18, a. m...							506	2.0												
18, a. m...	176	5.7	179	5.6	196	5.1	354	2.8												
20, a. m...							590	1.7	738	1.4	654	1.5								
20, a. m...	164	6.1	176	5.7	208	4.8	195	5.1	312	3.2										
21, p. m...	144	6.9	166	6.0	147	6.8														
21, p. m...					100	6.2														
22, a. m...			147	6.8	162	6.2	386	2.6	901	1.1	2,068	0.5								
22, p. m...	133	7.5																		
24, a. m...	203	4.9																		
24, a. m...	159	6.3																		
26, p. m...							233	4.3	296	3.4										
26, p. m...											298	3.4								
26, p. m...											231	4.3								
26, p. m...							183	5.5	186	5.4										
27, p. m...							439	2.3	285	3.5										
27, p. m...	136	7.4	139	7.2	151	6.6			161	6.2										
28, a. m...							106	6.0	193	5.2	185	5.4								
28, p. m...					141	7.1			190	5.3										
29, a. m...							1,147	0.9	306	2.5										
29, a. m...	140	7.1			216	4.6	272	3.7												
30, a. m...			152	6.6	183	5.5	222	4.5												
30, p. m...									181	5.5	202	5.0								
30, p. m...	113	8.8			151	6.6	139	7.2	167	6.0										
31, a. m...	137	7.3	172	5.8																
31, a. m...	134	7.5																		

Temperature—Continued.

CLOUDY.

Date.	1,000 feet. Rate.	Gradient.	1,500 feet. Rate.	Gradient.	2,000 feet. Rate.	Gradient.	3,000 feet. Rate.	Gradient.	4,000 feet. Rate.	Gradient.	5,000 feet. Rate.	Gradient.	6,000 feet. Rate.	Gradient.	7,000 feet. Rate.	Gradient.	8,000 feet. Rate.	Gradient.	Mean. Rate.	Gradient.
May 1, p.m...	272	3.7																		
4, p.m...	221	4.5																		
5, p.m...			216	4.6																
14, a.m...									375	2.7	286	3.5								
15, p.m...	148	6.8	170	5.9																
16, a.m...	587	5.3	397	2.5																
17, a.m...			488	2.0																
25, a.m...			943	1.1																
26, p.m...	120	8.3					215	4.7												
28, a.m...							1,558	0.6												
June 3, a.m...																				
3, a.m...									156	6.4	1,689	0.6								
										394	2.5								
5, p.m...	189	5.3	238	4.2	213	4.7														
6, p.m...			232	4.3																
6, p.m...			209	4.8																
8, p.m...	272	3.7																		
9, p.m...	144	6.9	183	7.5																
24, a.m...	152	6.6																		
25, a.m...	120	8.3																		
July 21, a.m...	147	6.8	167	6.0	171	5.8														
21, a.m...			130	7.7																
22, p.m...	442	2.3																		
31, a.m...	171	5.8	193	5.2	183	5.5														
31, a.m...			154	6.5	186	5.4														
Aug. 3, p.m...			144	6.9	169	5.9	189	5.3												
3, p.m...	108	9.3			124	8.1														
4, a.m...									803	8.2	548	1.8								
4, a.m...	127	7.9	187	6.0	258	3.9	360	2.8												
4, a.m...	93	10.8																		
7, a.m...	243	4.1	271	8.7	259	3.9	330	3.0	228	4.4										
7, a.m...	103	9.7	141	7.1			190	5.3												
17, a.m...	207	1.2																		
17, a.m...	221	4.5																		
17, a.m...	333	3.0																		
18, a.m...	2,086	0.5																		
18, a.m...	683	1.5																		
18, a.m...	238	4.2																		
Sept. 19, a.m...			127	7.3	173	5.8														
19, p.m...	162	6.2																		
Oct. 1, a.m...			1,451	0.7																
1, a.m...	236	4.2	165	6.1																
9, a.m...									167	6.0										
10, p.m...	139	7.2																		
12, p.m...	125	7.9																		
19, p.m...	183	5.5	195	5.1	205	4.9	292	3.4	314	3.2										
19, p.m...			1,071	0.9	469	2.1	360	2.8												
19, p.m...			165	6.1	173	5.8	186	5.4	196	5.1	220	4.5								
21, p.m...																				

SUMMARY.

CLEAR.

	1,000	1,500	2,000	3,000	4,000	5,000	6,000	7,000	8,000	Mean
Morning....	6.7	5.0	4.4	3.2	2.7	2.9	3.1	3.0		3.9
Afternoon...	6.3	5.8	5.4	5.2	4.4	3.6	3.4	3.5	4.9	4.7
Mean...	6.6	5.3	4.7	3.7	3.1	3.1	3.2	3.2	4.9	4.2

CLOUDY.

	1,000	1,500	2,000	3,000	4,000	5,000	6,000	7,000	8,000	Mean
Morning....	5.3	4.8	5.0	4.8	2.3					4.4
Afternoon...	6.0	5.0	5.2	3.4	4.8	4.5				5.8
Mean...	5.6	4.9	5.1	4.2	3.4	4.5				4.6

COMBINED.

	1,000	1,500	2,000	3,000	4,000	5,000	6,000	7,000	8,000	Mean
Morning....	6.4	5.0	4.5	3.3	2.6	2.9	3.1	3.0		3.8
Afternoon...	6.2	5.6	5.4	4.8	4.4	3.7	3.4	3.5	4.9	4.7
Mean...	6.3	5.2	4.8	3.7	3.1	3.2	3.2	3.2	4.9	4.2

Temperature—Continued.

INVERSIONS.

CLEAR.

Date.	800 feet. Rate.	Gradient.	1,000 feet. Rate.	Gradient.	1,200 feet. Rate.	Gradient.	1,500 feet. Rate.	Gradient.	2,000 feet. Rate.	Gradient.	2,500 feet. Rate.	Gradient.	3,000 feet. Rate.	Gradient.	3,500 feet. Rate.	Gradient.	4,000 feet. Rate.	Gradient.	4,500 feet. Rate.	Gradient.	5,000 feet. Rate.	Gradient.
May 8, a. m							666	1.5														
12, a. m																						
12, a. m											409	2.4										
28, a. m			432	2.3							2,739	0.4										
28, a. m											1,557	0.6										
30, a. m									395	2.5												
30, a. m									446	2.2												
31, a. m							522	1.9	515	1.9												
31, a. m							616	1.6														
June 21, a. m							289	3.5	469	2.1												
24, a. m							566	1.8														
24, a. m							844	1.2	978	1.0												
25, a. m													1,841	0.5								
29, a. m									459	2.2	2,168	0.5										
30, a. m											1,149	0.9										
July 6, a. m									567	1.8	3,330	0.3										
18, a. m					286	3.5	274	1.6	276	3.6												
18, a. m					197	5.1	439	2.3	336	3.0												
19, a. m											716	1.4	4,046	0.2								
22, 9 p. m							1,092	0.9														
27, a. m							156	6.4	222	4.5	472	2.1	357	2.8								
27, a. m							1,054	0.9	500	2.0	654	1.5	497	2.0								
Aug. 8, a. m			281	3.6			542	1.8	783	1.3												
13, a. m							662	1.5	1,723	0.6	926	1.1	365	2.7			1,055	0.9				
15, a. m							109	9.2	142	7.0	164	6.1										
15, a. m									221	4.5	278	3.6	496	2.0								
15, a. m					1,876	0.5			775	1.3	975	1.0	899	1.1								
16, a. m									228	4.4	588	1.7	379	2.6								
20, a. m											1,517	0.7										
22, a. m									287	3.5	488	2.0	1,619	0.6								
23, a. m									578	1.7	551	1.8	2,189	0.5								
28, a. m											463	2.2										
28, a. m											895	1.1	5,020	0.2								
30, a. m											332	3.0	676	1.5	1,573	0.6						
31, a. m											412	2.4	1,486	0.7								
Sept. 1, a. m											430	2.3	1,981	0.5								
2, a. m	347	2.9							386	2.6	920	1.1										
3, a. m															1,700	0.6						
4, a. m													471	2.1	2,783	0.4	6,407	0.2				
5, a. m									1,072	0.9			1,036	1.0								
7, 8 p. m							321	3.1														
17, a. m							216	4.6	268	3.7												
24, a. m							117	8.5	137	7.3	151	0.7										
24, a. m					207	4.8			245	4.1												
28, a. m													380	2.6	728	1.4	879	1.1				
29, a. m	150	6.7			192	5.2																
Oct. 2, a. m					2,570	0.4	1,245	0.8	1,357	0.7	418	2.4	2,787	0.4								
3, a. m					106	9.4	130	7.7														
4, 5 p. m											431	2.3	629	1.6	875	1.1						
5, a. m									1,262	0.8	777	1.3										
6, 7 p. m	385	2.6																				
11, 7 p. m	283	3.5																				
12, a. m									181	5.5	260	3.8										
14, a. m									227	4.4	564	1.8	1,299	0.8								
15, a. m													586	1.7								
18, a. m															765	1.3	966	1.0				
23, a. m																	374	2.7	466	2.1	451	2.2
29, a. m							216	4.6	478	2.1	646	1.5	1,969	0.5								
31, a. m									164	6.1	192	5.2										
31, a. m									1,988	0.5												

CLOUDY.

Date.	800 feet. Rate.	Gradient.	1,000 feet. Rate.	Gradient.	1,200 feet. Rate.	Gradient.	1,500 feet. Rate.	Gradient.	2,000 feet. Rate.	Gradient.	2,500 feet. Rate.	Gradient.	3,000 feet. Rate.	Gradient.	3,500 feet. Rate.	Gradient.	4,000 feet. Rate.	Gradient.	4,500 feet. Rate.	Gradient.	5,000 feet. Rate.	Gradient.
June 4, 8 p. m			514	1.9																		
Aug. 14, a. m					731	1.4	499	2.0	360	2.8	432	2.3	1,446	0.7								
14, a. m							1,322	0.8			2,465	0.4										
17, a. m			544	1.8	741	1.3																
Oct. 1, a. m							257	3.9	351	2.8			441	2.3								

DUBUQUE, IOWA.

TEMPERATURE.

Of the Dubuque series 148 observations at altitudes of 1,000 feet or more were examined. The number of ascensions was 65, and the greatest altitude attained, 5,065 feet.

The mean gradient was found to be 4.6° per thousand feet. There is a steady and quite uniform decrease from 1,000 up to 3,000 feet, and a further slight decrease up to 4,000 feet, followed by a slight increase thereafter. The greatest gradient was 6.9°, and the least, 3.2° per thousand feet. Up to 3,000, 4,000, and 5,000 feet they agree almost exactly with those at Omaha.

The average morning and afternoon and the mean gradients were exactly equal, but the two former varied irregularly up to the different elevations. This station was one of the four at which the afternoon gradient did not nearly equal or exceed the adiabatic rate, the other three being Dodge City, Springfield, and Duluth.

Except up to 1,000 feet, the clear-weather gradients were at all elevations considerably greater than the cloudy ones. The greatest difference occurred up to 1,500 feet, where the clear-weather gradient was 7°, and the cloudy one, 4.8° per thousand feet.

The cloud effects, when at all noticeable, were quite dissimilar to those previously experienced in the western country. Instead of a retardation of the temperature fall, the usual effect was an acceleration. Thus, on May 31 there was a fall of 7° from 2,400 to about 3,000 feet of elevation, the kite being in the clouds at the latter height, followed by a rise of 14° as the kite descended to the base of the low clouds at 1,000 feet elevation. On June 11 there was a fall of 7.5° as the kite disappeared in the clouds, with only about 100 feet increase in altitude, and a slow rise as the kite descended.

In the few instances in which the kite ascended above the clouds there was not the usual sharp rise at the place of emergence and a slower fall than usual thereafter, but, instead, a slow rise which in no case exceeded 4°, and which did not extend for more than 1,000 or 1,100 feet above the top of the clouds.

The inversions of temperature were infrequent, and presented no features of especial interest.

WINDS.

The wind directions conformed steadily to the usual rule. There were very few deflections toward the left, and none in either direction was at all marked.

RELATIVE HUMIDITY AND VAPOR PRESSURE.

The relative humidity changes with difference of altitude were very small, the greatest, 8 per cent, being found at 4,000 feet, and the least, 1 per cent, at 2,000 and 3,000 feet. From 2,000 to 4,000 feet there is a steady and regular decrease, followed by a sharp rise of 13 per cent at 5,000 feet. At all altitudes the humidities were higher than at the surface, except at 4,000 feet, at which height was also found the greatest departure from the surface humidity.

The effects of the presence of clouds upon the relative humidity were very much the same as at other stations. There would be a decided rise in most cases as the kite came near or into the clouds, at one time amounting to 33 per cent, followed by a fall as the kite was freed from cloud influence, but not to an equal extent.

The vapor pressure results agree almost exactly with those at Duluth, and maintain a steady relation with those at Dodge City, averaging 0.04 lower.

DIMINUTION OF VAPOR PRESSURE WITH ALTITUDE.

VALUE OF $\frac{p}{p^0}$ AT EACH RESPECTIVE 1,000 FEET OF ALTITUDE.

1,500 feet.	2,000 feet.	3,000 feet.	4,000 feet.	5,000 feet.
0.83	0.79	0.76	0.66	0.56

TEMPERATURE.

CLEAR.

Date.	1,000 feet.		1,500 feet.		2,000 feet.		3,000 feet.		4,000 feet.		5,000 feet.		Mean.	
	Rate.	Gradient.	Rate.	Gradient.	Rate.	Gradient.	Rate.	Gradient.	Rate.	Gradient.	Rate	Gradient.	Rate.	Gradient.
May 26, p.m	165	6.1												
June 2, p.m			176	5.7										
3, a.m					143	7.0	275	3.6						
3, p.m	174	5.7	130	7.7										
4, p.m	122	8.2	167	6.0	207	4.8								
10, p.m			177	5.0	157	6.4								
12, p.m	188	5.3	160	6.2										
18, a.m	152	5.6	139	7.2	246	4.1								
19, a.m	130	7.7	114	8.8										
20, a.m	143	7.0												
21, a.m					1,039	9.6								
21, a.m					191	5.2								
25, a.m					280	3.6	360	2.8						
27, a.m					163	6.1								
28, p.m			150	6.3										
July 5, p.m			124	8.1	145	6.9								
6, a.m					249	4.0	211	4.7						
6, a.m			142	7.0	117	8.5	163	6.1						
6, p.m			176	5.7										
17, a.m					253	4.0								
17, a.m					163	6.1								
18, a.m							509	2.0	263	3.8				
20, a.m							813	1.2	518	1.9	438	2.3		
20, a.m					181	5.5	272	3.7						
28, a.m			113	8.8										
Aug. 5, p.m					174	5.7	198	5.1						
5, p.m					190	5.3	185	5.4						
10, a.m					459	2.2	224	4.5						
11, a.m					199	5.0	209	4.8	183	5.5				
11, a.m			124	8.1	120	8.3	151	6.6						
12, a.m					150	6.7	164	6.1						
22, a.m			139	7.2										
Sept. 1, a.m							277	3.6	202	5.0	179	5.6		
2, p.m					878	2.6	316	3.2	257	3.9				
2, p.m							162	6.2	200	5.0	230	4.3		
19, a.m			134	7.5										
21, a.m							302	3.3	288	3.5				
21, p.m							226	4.4						
28, a.m					732	1.4	438	2.3						
28, p.m					237	4.2	280	3.6						
29, a.m							547	1.8	317	3.2	238	4.2		
29, p.m							189	5.3	220	4.5				
Oct. 24, a.m					1,039	1.0								
24, a.m					440	2.3								
26, a.m					194	5.2								
31, a.m			162	6.2										
31, a.m							445	2.2						
31, p.m									806	1.2	793	1.3		
31, p.m									1,316	0.8				

CLOUDY.

Date.	1,000 feet.		1,500 feet.		2,000 feet.		3,000 feet.		4,000 feet.		5,000 feet.		Mean.	
May 20, p.m	97	10.3												
27, a.m	212	4.7												
29, a.m	171	5.8	139	7.2										
31, a.m	133	7.5			114	8.8								
31, p.m			104	9.6										
June 1, a.m			171	5.8										
5, p.m	86	11.6	147	6.8										
9, p.m	272	3.7	252	4.0	237	4.2								
11, a.m	110	9.1			245	4.1								
14, a.m	157	6.4	181	5.5			331	3.0						
15, a.m							990	1.0						
15, a.m							400	2.5						
15, p.m			152	6.6			274	3.6						
22, a.m							374	2.7						
22, a.m			293	3.4	442	2.3								
23, a.m			375	2.7	452	2.2								
23, a.m					220	4.5								
24, a.m					1,065	0.9	440	2.3						
26, p.m	372	2.7												
30, a.m			958	1.0										
30, a.m			551	1.8										
Aug. 2, p.m	198	5.1	167	6.0										
6, a.m					999	1.0	293	3.4						
15, a.m			235	4.3										
16, a.m	134	7.5	236	4.2	1,102	0.9								
Sept. 3, a.m			576	1.7										
6, a.m							663	1.5	543	1.8	500	1.8		
6, a.m							213	4.7	301	3.3				
12, a.m	100	10.0												
14, a.m							309	2.7	318	3.1	306	3.3		

Date.	1,000 feet.		1,500 feet.		2,000 feet.		3,000 feet.		4,000 feet.		5,000 feet.		Mean.	
	Rate.	Gradient.	Rate.	Gradient.	Rate.	Gradient.	Rate.	Gradient.	Rate.	Gradient.	Rate.	Gradient.	Rate.	Gradient.
Oct. 5, a. m.					155	6.5								
6, p. m.			161	6.2										
7, a. m.					295	3.4								
9, p. m.					204	4.9								
16, a. m.					913	1.1								
22, a. m.					164	6.1								
23, p. m.							983	1.0						
27, p. m.							624	1.0	842	1.2				

SUMMARY.
CLEAR.

	1,000		1,500		2,000		3,000		4,000		5,000		Mean	
Morning		7.1		7.6		5.1		3.5		3.8		4.0		5.2
Afternoon		6.3		6.4		5.0		5.0		3.1		2.8		4.8
Mean		6.7		7.0		5.1		4.0		3.5		3.5		5.0

CLOUDY.

Morning		7.5		3.4		3.5		2.6		2.7		2.6		3.7
Afternoon		6.5		6.6		4.6		2.1		1.2				4.2
Mean		7.0		4.8		3.0		2.5		2.4		2.6		3.8

COMBINED.

Morning		7.4		5.4		4.4		3.2		3.5		3.4		4.6
Afternoon		6.4		6.5		4.9		4.2		2.8		2.8		4.6
Mean		6.9		5.9		4.6		3.5		3.2		3.3		4.6

NORTH PLATTE, NEBR.

TEMPERATURE.

In the North Platte observations 524 were considered at altitudes of 1,000 feet or more, and of these, 216 were taken in the morning and 308 in the afternoon. The highest altitude attained was 5,600 feet and the number of ascensions was 132.

The mean gradient was found to be 5.6° per thousand feet. This is much greater than those at Dodge City and Omaha. The rate up to 1,000 feet was 6.8°, and up to the remaining elevations as follows: Up to 1,500 feet, 6.5° per thousand feet; 2,000 feet, 5.9°; 3,000 feet, 5.2°; 4,000 feet, 4.4°, and 5,000 feet, 4.7°. (See plate 1.)

There is, as is usual in this district, a decrease up to 4,000 feet and an increase thereafter, and the rate of increase and decrease was quite regular.

The morning curve closely parallels the mean, while the afternoon one occupies a similar position on the opposite side of the mean, and at almost equal distances from it.

The clear-weather gradients were greater than the cloudy ones except up to 1,000 feet, the greatest difference being found up to 1,500 feet, where it was 1° per thousand feet.

The cloud effects were, as a rule, confined to a partial or complete suspension of the temperature fall. At times there was a rise, followed by a fall as soon as the kite emerged, and in about one-half the cases the clouds appeared to exercise little or no effect. None was sufficiently marked to be worthy of special note except on June 30, when a remarkable inversion due to clouds took place. The sky was partially overcast with clouds, and fresh east to east-southeast winds were blowing at the surface, inclining somewhat to the right above, with moderate velocities, as indicated by the pull on the kite wire. At 600 feet elevation there was one-half degree rise in the temperature, but at 1,500 feet the excess amounted to 5°, and steadily increased to 13.5° at 3,000 feet. The temperature at the kite rose 17.5° with increase of height to 3,000 feet, while that at the surface rose but 3.5° during the same time, which was from 7.25 to 8.45 a. m. The day previous had been quite warm, and the upper air had become warmed to a considerable extent.

The clear night had promoted surface radiation, and the early cloud formation, by cutting off the sun's heat in the morning, prevented the usual readjustment. The cloud dissipated about 8.45 a. m., and then occurred a sudden rise in the surface temperature of 12°, and a gradual return thereafter to somewhat normal conditions.

The inversions were comparatively few and without special characteristics except in the one instance of June 30, which has just been described. On August 4 there was a rise of 12.5° between 1,900 and 2,600 feet as the kite emerged above a cloud, accompanied by a fall in the humidity of 30 per cent. As soon as the cloud had moved away the temperature fell 10°, and the humidity rose 21 per cent, although the kite had fallen 400 feet.

The wind direction and velocity appeared to exercise no effect. It is true that the gradients were, as a rule, much greater with west, particularly northwest, than with east winds, but the east winds were generally associated with cloudy weather, and the decreased gradients at these times must be attributed to the presence of the clouds.

WINDS.

Wind directions followed the usual rule of deflection toward the right, and to about the same extent as at Omaha, but less than at Dodge City. Movements in the opposite direction were infrequent and not decided, and in a majority of cases were followed by rain and thunderstorms.

RELATIVE HUMIDITY AND VAPOR PRESSURE.

The relative humidity conformed to the general rule of steady decrease with increase of altitude, but there is a departure from the rule in that the humidity above is uniformly greater than that at the ground, being 3 per cent greater at 1,500 feet, and increasing slowly to 7 per cent at 5,000 feet. It is suggested that this increase is due to the preponderance of west to north winds, which, with their lower temperatures, cause a rise in the relative humidity, the quantity of moisture in the atmosphere remaining approximately the same.

Cloud effects were indifferent and irregular. In a majority of cases the proximity of rain produced no effect either way. Occasionally there would be a rise previous to the commencement of rain, at one time amounting to 28 per cent, but such cases were very rare. The presence of the kite in the clouds was also without event, except in the one case of August 4. As a rule the only effect was to maintain the humidity at a nearly stationary point, without regard to altitude or time of day. On August 4, as noted before, as the kite emerged above a cloud there was a fall of 30 per cent in the humidity, due to a rise in the temperature of 12.5°, and a rise of 21 per cent as the kite descended after the cloud had disappeared, the temperature falling 10° at the same time.

The vapor pressure results are given below. It will be noticed that they agree almost exactly with the Omaha results between 3,000 and 5,000 feet.

DIMINUTION OF VAPOR PRESSURE WITH ALTITUDE.

VALUE OF $\frac{p}{p_0}$ FOR EACH RESPECTIVE 1,000 FEET OF ALTITUDE.

1,500 feet.	2,000 feet.	3,000 feet.	4,000 feet.	5,000 feet.
0.78	0.72	0.66	0.57	0.40

TEMPERATURE.

CLEAR.

Date.	1,000 feet.		1,500 feet.		2,000 feet.		3,000 feet.		4,000 feet.		5,000 feet.		Mean.	
	Rate.	Gradient.	Rate.	Gradient.	Rate.	Gradient.	Rate.	Gradient.	Rate.	Gradient.	Rate.	Gradient.	Rate.	Gradient.
May 7, a. m.....			95	10.5	103	9.7								
7, p. m.....			89	14.5	115	8.7	189	7.2	162	6.2				
8, p. m.....			83	12.0	91	11.0	113	8.8	138	7.2	152	6.6		
10, a. m.....			105	9.5	154	6.5	137	7.3						
10, p. m.....			116	8.6			142	7.0						
11, a. m.....			216	4.6	266	3.8	210	4.8	197	5.1	180	5.6		
11, p. m.....									174	5.7				
18, p. m.....			133	7.5	161	6.2								
21, a. m.....	276	3.6	404	2.5										
21, p. m.....	154	6.5	149	6.7	170	5.9	163	6.1	158	6.3				
23, a. m.....			438	2.3	334	3.0	294	3.4						
29, a. m.....	140	7.1	404	2.5	271	3.7								
30, p. m.....			360	2.8										
June 1, a. m.....	149	6.7												
1, p. m.....			140	7.1	157	6.4	263	3.8						
2, p. m.....	187	5.3	248	4.0										
3, p. m.....	67	14.9	139	7.2	143	7.0	160	6.2						
8, a. m.....	123	8.1												
12, p. m.....	187	5.3												
15, p. m.....			438	2.3										
19, p. m.....	106	9.4	110	9.1	106	9.4	130	7.7						
20, a. m.....			113	8.8	129	7.8								
21, p. m.....	500	2.0												
23, p. m.....	139	7.2	173	5.8	166	6.0	214	4.7	200	5.0	207	2.5		
23, a. m.....									771	1.3				
23, a. m.....			132	7.5	163	6.1	217	4.6	349	2.9				
2?, p. m.....	138	7.2	146	6.8	152	6.6	173	5.8						
26, a. m.....	1,658	0.6	622	1.6	170	5.9	208	4.8						
26, a. m.....			164	6.1										
27, p. m.....			139	7.2	143	7.0								
29, a. m.....	273	3.7	229	4.4	263	3.7								
July 1, a. m.....	126	7.0												
2, a. m.....	236	4.2												
4, p. m.....	216	4.6	141	7.1	135	7.4	171	5.8						
5, a. m.....	121	8.3	160	6.2	157	6.4								
5, a. m.....	134	7.5												
10, p. m.....	130	7.7	117	8.5										
11, a. m.....	106	9.4	194	5.2	225	4.4								
12, p. m.....	126	7.9	133	7.5	194	5.2	171	5.8						
13, a. m.....			228	4.4	249	4.0	240	4.0	202	5.0				
13, p. m.....			106	9.4	124	8.1	149	6.7						
14, a. m.....	263	3.8	142	7.0	188	5.3								
15, a. m.....	127	7.9	129	7.8										
16, p. m.....			117	8.5	124	8.1								
17, a. m.....	136	7.4			525	1.9	389	2.6	277	3.6				
17, p. m.....			121	8.3	236	4.2	321	3.1						
18, p. m.....	103	9.7	167	6.0	178	5.6	176	5.7						
19, a. m.....			172	5.8	187	5.3	265	3.8						
19, a. m.....	184	5.4	213	4.7										
20, p. m.....			118	8.5	139	7.2	185	6.1						
21, a. m.....			973	1.0	815	1.2	343	2.9						
21, p. m.....			118	8.5	135	7.4	202	5.0						
22, p. m.....	127	7.9	134	7.5	142	7.0	161	6.2						
23, p. m.....			153	6.5	165	6.1	148	6.8						
24, a. m.....	106	9.4	141	7.1	150	6.7								
26, p. m.....	132	7.6												
26, p. m.....			146	6.8	139	7.2	174	5.7						
27, a. m.....			172	5.8	190	5.3	235	4.3						
27, p. m.....			132	7.6	145	6.9	168	6.0						
28, a. m.....			161	6.2	190	5.3	177	5.6						
28, p. m.....					129	7.8	180	5.6						
29, a. m.....			160	6.2										
Aug. 3, a. m.....			180	5.6	141	7.1	197	5.1						
3, p. m.....			120	8.3										
5, a. m.....	84	11.9	418	2.4	406	2.5	391	2.6						
9, a. m.....	106	9.4	135	7.4	175	5.7	197	5.1	201	5.0				
15, p. m.....	120	8.3	143	7.0	143	7.0								
16, a. m.....			350	2.9	513	1.9	858	1.2						
16, p. m.....									286	3.5				
17, p. m.....	127	7.9	159	6.3	164	6.5								
18, p. m.....	174	5.6												
19, a. m.....			156	6.4										
20, p. m.....	2,222	0.5	537	1.9	276	3.6	251	4.0						
21, p. m.....			161	6.2	154	6.5	175	5.7						
21, p. m.....			144	6.9	174	5.7								
26, p. m.....	94	10.6												
27, a. m.....									1,324	0.8				
27, a. m.....			95	10.6	117	8.5	141	7.1	190	5.3				
28, a. m.....	100	10.0	187	7.3	134	7.5	169	5.9						
28, p. m.....			102	9.8	112	8.9								
29, p. m.....	90	11.1	145	6.9	132	7.6	154	6.5						
30, a. m.....			153	6.5	199	5.0	202	5.0	247	4.0				
30, p. m.....			116	8.6	127	7.9	139	7.2						
Sept. 1, a. m.....	109	9.2			938	1.1	150	6.7						
1, a. m.....			182	7.6	127	7.3								
2, p. m.....	117	8.5	135	7.4	135	7.4								

Temperature—Clear—Continued.

Date.	1,000 feet.		1,500 feet.		2,000 feet.		3,000 feet.		4.000 feet.		5,000 feet.		Mean.	
	Rate.	Gradient.	Rate.	Gradient.	Rate.	Gradient.	Rate.	Gradient.	Rate.	Gradient.	Rate.	Gradient.	Rate.	Gradient.
Sept. 4, p. m.			91	11. 0	115	8. 7	140	7. 1						
4, p. m.			97	10. 3	86	11. 6								
6, a. m.			161	6. 2	157	6. 4	170	5. 9	176	5. 7				
6, p. m.			80	12. 5	100	10. 0	114	8. 8						
7, p. m.	81	12. 3	90	10. 1	125	8. 8	143	6. 9						
8, a. m.							943	1. 1						
8, p. m.			92	10. 9			118	8. 5						
17, p. m.			130	7. 7	139	7. 2	157	6. 4						
19, p. m.	151	6. 6	170	5. 9	182	5. 5								
19, p. m.			148	6. 8										
20, a. m.							651	1. 5	516	1. 9	243	4. 1		
20, p. m.					134	7. 5	184	5. 4	190	5. 3				
21, a. m.	189	5. 3	169	5. 9										
22, p. m.	185	5. 4	191	5. 3	199	5. 0								
22, p. m.	291	3. 4												
23, p. m.	193	5. 2	214	4. 7	222	4. 5	201	5. 0						
23, p. m.			269	3. 7										
24, p. m.	117	8. 5	137	7. 3										
26, p. m.	125	8. 0	128	7. 8										
27, a. m.			135	7. 4	478	2. 1	360	2. 8	402	2. 5	291	3. 4		
27, p. m.							222	4. 5	170	5. 9	227	4. 4		
29, p. m.	353	2. 8												
30, a. m.			144	6. 9	266	3. 8	149	6. 7						
30, p. m.			86	11. 6	107	9. 3								
Oct. 2, a. m.			116	8. 6	141	7. 1	144	6. 9	150	6. 7	198	5. 1		
2, p. m.			97	10. 3	123	8. 1	124	8. 1						
4, p. m.			140	7. 1	122	8. 2	139	7. 2	163	6. 1				
4, p. m.			150	6. 7	156	6. 4	168	6. 0						
10, a. m.			149	6. 7	196	5. 1	226	4. 4	195	5. 1				
12, a. m.			232	4. 3										
12, a. m.			122	8. 2										
13, a. m.	117	8. 5	107	9. 3	124	8. 1	167	6. 0						
14, p. m.			256	3. 9	211	4. 7	263	3. 4						
14, p. m.					763	1. 3								
18, p. m.			206	4. 9	162	6. 2	184	5. 4	307	3. 3				
18, p. m.			382	2. 6	230	4. 3	268	3. 7						
10, p. m.			479	2. 1	408	2. 5	405	2. 5						
21, p. m.	176	5. 7	155	6. 5	125	8. 0								
21, p. m.			87	11. 5										
23, p. m.			229	4. 4	199	5. 0	219	4. 6	259	3. 9				
23, a. m.									349	2. 9				
25, a. m.			146	6. 8	161	6. 2	214	4. 7	249	4. 0	226	4. 4		
25, p. m.			83	12. 0	80	12. 5	137	7. 3	159	6. 3	197	5. 1		
26, p. m.			108	9. 3	127	7. 9	176	5. 7	187	5. 3				
27, p. m.			129	7. 8	147	6. 8	146	6. 8	183	5. 5	187	5. 3		
27, p. m.			96	11. 1	99	10. 1	129	7. 8	147	6. 8	163	6. 1		
28, a. m.			152	6. 6	175	5. 7								
28, p. m.							222	4. 5	207	4. 8				
28, p. m.			104	9. 6	121	8. 3	141	7. 1	174	5. 7				
29, p. m.			111	9. 0	171	5. 8								
29, p. m.			157	6. 4										
29, p. m.			172	5. 8										
30, p. m.			114	8. 8	157	6. 4	159	6. 3	292	3. 4				
30, p. m.			398	2. 5	266	3. 8	256	3. 9						

CLOUDY.

Date.	Rate.	Gradient.	Rate.	Gradient.	Rate.	Gradient.	Rate.	Gradient.	Rate.	Gradient.	Rate.	Gradient.	Rate.	Gradient.
May 1, a. m.			239	4. 2	245	4. 1	304	3. 3						
1, p. m.			133	7. 5	141	7. 1								
2, p. m.			105	9. 5										
3, p. m.			149	6. 7	202	5. 0	209	4. 8						
4, p. m.			185	5. 4	216	4. 6	265	3. 8						
5, a. m.	242	4. 1	290	3. 4	338	3. 0								
5, p. m.			135	7. 4	152	6. 6	167	6. 0						
12, p. m.			203	4. 9	123	8. 1	150	6. 7						
13, a. m.			180	5. 6	192	5. 2	238	4. 2	343	2. 9				
13, p. m.							171	5. 8						
14, p. m.			146	6. 8										
15, a. m.	310	3. 2	404	2. 5										
15, p. m.	280	3. 6	373	2. 7	245	4. 1								
16, p. m.			190	5. 3	201	5. 0	220	4. 5						
17, a. m.			210	4. 8	189	5. 3	163	6. 1	232	4. 3	252	4. 0		
20, a. m.	258	3. 9												
23, a. m.			252	4. 0	321	4. 1	210	4. 8						
24, a. m.							1, 258	0. 8						
24, p. m.	205	4. 9	180	5. 6										
25, a. m.	185	5. 4	214	4. 7										
26, a. m.	126	7. 9	146	6. 8	184	5. 4								
27, a. m.	224	4. 5	193	5. 3	242	4. 1	224	4. 5						
28, a. m.	682	1. 5	1, 259	0. 8										
28, p. m.	146	6. 8	182	5. 5										
31, a. m.	180	5. 6	157	6. 4										
June 4, p. m.	212	4. 7	308	3. 2	287	3. 2								
9, p. m.			283	3. 5	302	3. 3								
10, a. m.	248	4. 0	173	5. 8	252	4. 0								
14, a. m.	809	1. 2	1, 021	1. 0	1, 364	0. 7								
14, p. m.			234	4. 3	217	4. 6								
30, a. m.			130	7. 7	152	6. 6	785	1. 4	1, 208	0. 8				

*Temperature—Cloudy—*Continued.

Date.	1,000 feet.		1,500 feet.		2,000 feet.		3,000 feet.		4,000 feet.		5,000 feet.		Mean.	
	Rate.	Gradient.	Rate.	Gradient.	Rate.	Gradient.	Rate.	Gradient.	Rate.	Gradient.	Rate.	Gradient.	Rate.	Gradient.
July 8, p. m	93	10.8	175	5.7	142	7.0								
9, a. m....			278	3.6	469	2.1	299	3.3						
9, p. m			143	7.0	148	6.8								
11, p. m			156	6.4	129	7.8								
Aug. 1, a. m....			209	4.8	228	4.4	191	5.2						
4, a. m....			183	5.5	154	6.5								
4, a. m....			96	10.4										
23, a. m....			255	3.9	239	4.2								
23, p. m			121	2.4										
Sept. 10, p. m			206	4.9	313	3.2	290	3.4	306	3.3				
10, p. m			228	4.4	202	5.0	300	3.3						
16, p. m	71	14.1	108	9.3	119	8.4								
Oct. 3, a. m....			215	4.7	830	1.2	663	1.5	550	1.8				
3, p. m			110	9.1	139	7.2	215	4.7	405	2.5				
9, a. m....			279	3.6										
9, p. m			125	7.9										
15, p. m			164	6.1	155	6.5	167	6.0	212	4.7	239	4.2		
15, p. m			251	4.0	202	5.0	224	4.5	215	4.7				
16, a. m....			232	4.3	176	5.7	204	4.9	190	5.3				
16, p. m			112	8.0	153	6.5	164	6.1						
20, a. m....	100	10.0	150	6.7	180	5.6	250	4.0	270	3.7				
24, a. m....	110	9.1	166	6.0	202	5.0	347	2.9						

SUMMARY.

CLEAR.

	1,000		1,500		2,000		3,000		4,000		5,000		Mean	
Morning..........		6.8		6.2		5.0		4.5		3.9		4.2		5.1
Afternoon........		7.3		7.3		7.0		5.9		5.3		5.5		6.7
Mean........		7.1		6.9		6.3		5.4		4.6		4.8		5.8

CLOUDY.

	1,000		1,500		2,000		3,000		4,000		5,000		Mean	
Morning..........		5.0		4.9		4.6		3.5		3.1		4.0		4.2
Afternoon........		7.5		5.9		5.4		5.0		3.8		4.2		5.3
Mean........		5.8		5.4		5.1		4.3		3.4		4.1		4.7

COMBINED.

	1,000		1,500		2,000		3,000		4,000		5,000		Mean	
Morning..........		6.2		5.7		4.8		4.2		3.7		4.2		4.8
Afternoon........		7.3		7.0		6.6		5.7		5.0		5.3		6.0
Mean........		6.8		6.5		5.9		5.2		4.4		4.7		5.6

INVERSIONS.

Date.	600 feet.		800 feet.		1,000 feet.		1,200 feet.		1,500 feet.		2,000 feet.		2,500 feet.		3,000 feet.		3,500 feet.	
	Rate.	Gradient.	Rate.	Gradient.	Rate.	Gradient.	Rate.	Gradient.	Rate.	Gradient.	Rate.	Gradient.	Rate.	Gradient.	Rate.	Gradient.	Rate.	Gradient.
June 30..........									302	3.3	195	5.1	253	4.0	223	4.5		
Aug. 12, 9 p. m ..			1,638	0.6			2,816	0.4										
27..........									118	8.5	141	7.1	190	5.3	263	3.8	317	3.2
Sept. 1..........									941	1.1								
8..........									1,532	0.7	4,152	0.2						
20..........									525	1.9								
Oct. 14..........									1,061	0.9								
23..........									141	7.1	306	3.3			630	1.6		

OMAHA, NEBR.

TEMPERATURE.

There were considered of the Omaha series of observations 185, obtained from 61 ascensions, and the greatest altitude attained was 7,243 feet.

The general mean decrease of temperature with increase of altitude was at the rate of 4.1° for each 1,000 feet. The average decrease up to 1,500 feet was 5.4°; up to the other elevations it was as follows: up to 2,000 feet, 4.9° per thousand feet; 3,000 feet, 3.6°; 4,000 feet, 3.2°; 5,000 feet, 3.5°; 6,000 feet, 3.8°, and 7,000 feet, 4.1°. (See plate 1.) There is the usual decrease up to 4,000 feet, and a steady increase thereafter. The change is most rapid between 2,000 and 3,000 feet.

The morning gradients averaged 3.6° per thousand feet, 0.5° less than the mean. The curve fairly parallels the mean curve, the greatest difference being found at 2,000 and the least at 6,000 feet. The afternoon mean was 5.9° per thousand feet, and from 4,000 feet upward is very close to the true adiabatic rate, averaging 5.3° per thousand feet. The greatest gradient was up to 1,500 feet, where it was 7.1° per thousand feet. The clear-weather gradients were naturally greater than the cloudy ones, particularly up to 2,000 feet, where the difference was 1.4° per thousand feet. An exception occurs up to 5,000 feet; but as there was but a single observation made at that height during cloudy weather, a fair comparison can not be had.

Cloud effects were much the same as at other stations, though not in so marked a degree as at some. There would be a sudden check in the temperature fall, and often a change to a rise, amounting at one time, August 19, to 5° between 1,500 and 2,200 feet of altitude. Other slight rises were noted, and there were still other cases in which the clouds, while not of sufficient influence to cause a rise in the temperature, would hold it steady until the kite emerged. No cases of inversion due to clouds were noticed.

The inversions of temperature were very few, on account of the time of day at which the major portion of the observations were taken.

WIND.

The wind directions conformed to the usual rule, being practically alike above and at the surface. There was the usual deflection of the kite toward the right, but to a lesser degree than at Dodge City or Washington. There were very few deflections in the opposite direction, and none was at all marked.

RELATIVE HUMIDITY AND VAPOR PRESSURE.

The change in the relative humidity with elevation was quite decided, a steady decrease being the rule up to the highest elevations reached. At 1,500 feet the relative humidity was 4 per cent higher than at the ground; at 2,000 feet it was 1 per cent lower; at 4,000 feet, 12 per cent lower; at 5,000 feet, 19 per cent lower; at 6,000 feet, 28 per cent lower, and at 7,000 feet, 34 per cent lower.

Clouds and the close proximity of rain, either before or after the ascensions, exercised a very pronounced effect upon the humidity, the rise being rapid and decisive. At times it amounted to as much as 30 or 40 per cent, and on one day to 51 per cent. On this day, September 20, there was a belt of cloud at an altitude of about 4,500 feet, and as the kite reached it from above there was a rise in the humidity from 35 to 86 per cent, followed by a steady fall after the kite emerged.

The vapor-pressure results follow. It will be noticed that there is a rapid diminution above 2,000 feet, corresponding to the marked decrease in the relative humidity with increase of altitude.

DIMINUTION OF VAPOR PRESSURE WITH ALTITUDE.

VALUE OF $\frac{p}{p^0}$ FOR EACH RESPECTIVE 1,000 FEET OF ALTITUDE.

1,500 feet.	2,000 feet.	3,000 feet.	4,000 feet.	5,000 feet.	6,000 feet.	7,000 feet.
0.83	0.80	0.68	0.56	0.39	0.23	0.15

TEMPERATURE.

CLEAR.

Date.	1,000 feet.		1,500 feet.		2,000 feet.		3,000 feet.		4,000 feet.		5,000 feet.		6,000 feet.		7,000 feet.		Mean.	
	Rate.	Gradient.	Rate.	Gradient.	Rate.	Gradient.	Rate.	Gradient.	Rate.	Gradient.	Rate.	Gradient.	Rate.	Gradient.	Rate.	Gradient.	Rate.	Gradient.
June 17, a. m...			1,021	1.0	259	3.9												
17, a. m...			475	2.1														
18, p. m...			123	8.1	151	6.6												
20, a. m...					700	1.4	300	3.3										
22, a. m...									175	5.7								
23, a. m...									1,344	0.7			707	1.4				
23, a. m...													313	3.2				
23, p. m...							183	5.5	180	5.6								
24, a. m...					145	6.9	485	2.1										
26, a. m...					2,046	0.5	503	1.7										
26, a. m...					585	1.7												
27, a. m...									917	1.1	498	2.0						
28, a. m...					628	1.6			309	3.2								
29, a. m...					176	5.7	474	2.1	383	2.6								
Aug. 4, a. m...					653	1.5	257	3.9	259	3.9								
4, a. m...			123	8.1	134	7.5	190	5.3										
6, p. m...					179	5.6												
6, p. m...					140	7.1												
8, a. m...			660	1.5	211	4.7												
9, a. m...							214	4.7	218	4.5								
9, a. m...					146	6.8	211	4.7										
10, a. m...							833	1.2	1,073	0.9	1,003	1.0	436	2.3				
10, a. m...									191	5.2	293	3.4						
11, a. m...			177	5.6	178	5.6	244	4.1	253	4.0								
14, p. m...					231	4.3	229	4.4										
15, a. m...			128	7.8	142	7.0												
17, a. m...					194	5.2												
19, p. m...			119	8.4	129	7.8												
22, a. m...							1,582	0.8										
22, a. m...					151	6.6	180	5.6										
24, a. m...			155	6.5	131	7.6												
28, a. m...			155	6.5	217	4.6	375	2.7	339	2.9								
30, a. m...			245	4.1	520	1.9												
31, a. m...			125	8.0	126	7.9	141	7.1	619	1.6	216	4.6	307	3.3				
Sept. 1, a. m...							491	2.0	445	2.2	301	3.3	212	4.7				
1, p. m...									159	6.3	174	5.7						
2, a. m...									988	1.0	501	2.0						
2, a. m...											226	4.4						
3, a. m...					163	6.1												
4, a. m...									389	2.6								
4, p. m...			119	8.4	118	8.5	141	7.1										
5, a. m...							443	2.3	252	4.0								
6, p. m...					118	8.5			1,140	0.9	165	6.1						
7, a. m...									175	5.7	587	1.7						
7, p. m...			163	6.1	136	7.4	206	4.9										
8, p. m...							371	2.7	425	2.4	423	2.4						
18, a. m...			226	4.4	202	5.0	193	5.2										
18, p. m...							165	6.1	303	3.3	213	4.7	246	4.1	205	4.9		
20, a. m...													148	5.3				
20, p. m...					138	7.2												
20, p. m...									782	1.3	521	1.9	221	4.5				
21, a. m...									134	7.5	193	5.2						
21, p. m...											495	2.0	482	2.1	362	2.8	303	3.3
23, a. m...											208	4.8	321	3.1				
23, a. m...			260	3.8	250	4.0	1,003	1.0										
24, p. m...							288	3.5										
24, a. m...			276	3.6	250	4.0												
27, a. m...					729	1.4												
27, p. m...					185	5.4												
28, p. m...									201	5.0	201	5.0						
28, p. m...			88	11.4					162	6.2								
Oct. 5, a. m...					146	6.8	446	2.2	1,964	0.5								
13, a. m...					159	6.3	669	1.5	493	2.0								

CLOUDY.

Date.	1,000 feet.		1,500 feet.		2,000 feet.		3,000 feet.		4,000 feet.		5,000 feet.		6,000 feet.		7,000 feet.		Mean.	
June 14, a. m...					329	3.0	100	5.3										
21, a. m...			559	1.8	500	2.0	794	1.4										
Aug. 2, p. m...			162	6.2	192	5.2	314	3.2										
5, a. m...			142	7.0	193	5.2												
19, a. m...							421	2.4										
20, a. m...					657	1.5												
23, p. m...			127	7.9			390	2.6										
Sept. 5, p. m...					193	5.2	215	4.7	250	4.0								
6, a. m...					673	1.5	456	2.2										
9, a. m...					216	4.7	266	3.8	297	3.4								
9, a. m...			274	3.6			452	2.2										
11, a. m...							453	2.2	363	2.8								
11, a. m...							178	5.6										
13, p. m...					132	7.6	157	6.4										

Temperature—Cloudy—Continued.

Date.	1,000 feet.		1,500 feet.		2.000 feet.		3,000 feet.		4,000 feet.		5,000 feet.		6,000 feet.		7,000 feet.		Mean.	
	Rate.	Gradient.	Rate.	Gradient.	Rate.	Gradient.	Rate.	Gradient.	Rate.	Gradient.	Rate.	Gradient.	Rate.	Gradient.	Rate.	Gradient.	Rate.	Gradient.
Sept. 14, a. m...					359	2.8	366	2.7										
16, a. m...			297	3.4	331	3.0	385	2.6	807	1.2								
20, p. m...											158	6.3						
Oct. 19, a. m...			215	4.7	262	3.8	291	3.4										
19, a. m...					197	5.1	228	4.4										
20, a. m...					277	3.6												
21, a. m...			164	6.1														
27, a. m...			613	1.6														
29, a. m...			185	6.1														
30, a. m...			256	3.9														

SUMMARY.
CLEAR.

Morning...			4.8		4.6		3.0		2.6		2.9		3.4		4.1		3.6	
Afternoon...			7.1		6.5		6.9		5.5		5.0		5.3				6.0	
Mean...			5.9		5.3		3.7		3.3		3.4		3.8		4.1		4.2	

CLOUDY.

Morning...			4.2		3.3		3.2		2.5								3.3	
Afternoon...			7.0		6.0		4.6		4.0		6.3						5.6	
Mean...			4.8		3.9		3.4		2.8		6.3						4.2	

COMBINED.

Morning...			4.6		4.2		3.1		2.6		2.9		3.4		4.1		3.6	
Afternoon...			7.1		6.4		6.0		5.4		5.2		5.8				5.9	
Mean...			5.4		4.9		3.6		3.2		3.5		3.8		4.1		4.1	

PIERRE, S. DAK.

TEMPERATURE.

Four hundred and sixteen observations at altitudes of 1,000 feet or more were considered in computing the temperature gradients at Pierre; the number of ascensions was 134, and the highest altitude attained 6,059 feet. The mean gradient was found to be 4.6° for each thousand feet of ascent, 0.5° greater than that at Omaha, 1.0° less than that at North Platte, and the same as that at Dubuque.

Contrary to the usual experience at stations in this district, the decrease with increase of elevation disappears above 3,000 feet instead of above 4,000 feet, and is again resumed above the latter elevation, the minimum gradient being found up to 6,000 feet.

The inversions of temperature were for the most part confined to the usual conditions in such cases, namely, clear mornings and the radiation effects resulting therefrom. None were decidedly marked except in the month of September. On the 20th of this month there was an inversion of 16° at an elevation of 1,677 feet, or at the rate of 9.5° per thousand feet, at 8.16 a. m., and at 3,500 feet at 8.37 a. m. there was still an inversion of 6°, or at the rate of 1.7° per thousand feet. It did not entirely disappear until after the 4,000 foot level had been passed at 9.03 a. m. On September 23, during a cloudy morning, there was an inversion of 15.4° at an elevation of 3,050 feet, or at the rate of 5° per thousand feet, at 7.54 a. m., the excess above the surface temperature increasing with the altitude up to this point. The inversion was still 11.3° at 3,400 feet, or at the rate of 3.3° per thousand feet, and did not finally disappear until about 10 a. m., after the kite had passed up beyond the 4,000-foot level.

These inversions, of which twenty-four instances were found, occurred with but three exceptions on days when a low barometric area was central to the northwestward, causing warm southerly winds which were usually brisk.

More than two-thirds of the observations were taken in the morning, and the average morning gradient was 4.3° per thousand feet, 0.3° less than the mean, and 0.7° greater than that at Omaha. There are no great divergences from the mean curve, the maximum difference, 0.6° per thousand feet, occurring up to 1,000 feet.

The afternoon average was 6.3° for each thousand feet, 0.9° greater than the adiabatic rate. The rate of decrease was greater at all elevations than the adiabatic rate, and was 2.1° greater up to 1,000 feet. The clear weather gradients were greater than the cloudy ones, the mean difference being 0.9° per thousand feet. The afternoon means show the least difference. The greatest difference occurred up to 4,000 feet, where it was 2.6°, and the least up to 1,500 feet, where it was 0.5° per thousand feet. In both clear and cloudy weather the greatest gradients were found up to 1,000 feet, and the least up to 4,000 feet. An extremely large gradient was noted during the afternoon of September 26, when it was 13° up to 1,000 feet, and 10.5° per thousand feet up to 1,500 feet.

The clouds at times appeared to exercise considerable effect, and at others little or none. The usual result was a retardation of the temperature fall. On one date there was a fall of 4° as the kite reached the clouds, the elevation remaining practically the same while the temperature was falling. On June 22 there was a rise of 12° as the kite emerged above the clouds, and a much slower rate of fall thereafter, amounting to but 5.7° in 2,600 feet of ascent. There is evidence of a similar action on September 30 at 1,900 feet elevation, although no reference to clouds can be found. On October 2 the temperature commenced to rise when the clouds were reached, and rose 8° while the kite ascended from 1,700 to 2,700 feet. From thence up to 5,000 feet the usual decrease took place.

WINDS.

There were very few cases of kite deflection toward the left. Only two were noticeable to any extent, and rain fell shortly afterwards in both instances. The amount of deflection toward the right was also less than was usually observed, the directions above and at the surface being practically the same.

RELATIVE HUMIDITY AND VAPOR PRESSURE.

The changes with altitude in the relative humidity were very slight. In fact, it may be said that there was practically none. The extreme range was but 8 per cent, the highest humidity, 66 per cent, being found at 3,000 feet, and the lowest, 58 per cent, at 5,000 feet. The percentage at 6,000 feet was 64, but it is very probable that these comparatively high figures are due to the scarcity of observations at that level and that a greater number would have shown a decrease in the mean.

As at North Platte, the upper-air humidities were uniformly greater than those at the surface, the greatest differences being 15 per cent at 5,000 feet and 16 per cent at 6,000 feet, and the least, 5 per cent at 2,000 feet.

The preponderance of easterly winds, with their greater moisture content, may have some bearing upon this fact. Nearly 52 per cent of the winds were from some easterly direction, and 30 per cent were from the southeast.

The proximity of rain caused a rise in the humidity, as a rule, to some point above 80 or 85 per cent, the surface humidity showing a less rise, and in some instances a fall. During the summer months low humidities of course prevailed, and rain did not exercise so great an influence.

The clouds quite frequently caused a rise to total saturation as the kite reached them. At other times their effect was not so marked, but at all times was greater than at the remaining stations in this district.

The vapor pressure results, as given below, show a greater agreement between the amount of moisture above and at the surface than is usual, and correspond closely to the slight differences in the relative humidity before noted. They are greater at all altitudes than those at North Platte and Omaha, and are even greater than those at Dodge City, except at 3,000 feet, where there is a difference of 5 per cent in favor of the latter.

DIMINUTION OF VAPOR PRESSURE WITH ALTITUDE.

VALUE OF $\frac{p}{p^5}$ AT EACH RESPECTIVE 1,000 FEET OF ALTITUDE.

1,500 feet.	2,000 feet.	3,000 feet.	4,000 feet.	5,000 feet.	6,000 feet.
0.90	0.86	0.75	0.72	0.69	0.69

TEMPERATURE.

CLEAR.

Date.	1,000 feet.		1,500 feet.		2,000 feet.		3,000 feet.		4,000 feet.		5,000 feet.		6,000 feet.		Mean.	
	Rate.	Gradient.	Rate.	Gradient.	Rate.	Gradient.	Rate.	Gradient.	Rate.	Gradient.	Rate.	Gradient.	Rate.	Gradient.	Rate.	Gradient.
May 15, p. m ...			146	6.8	168	6.0										
23, p. m ...			162	6.2	147	6.8	227	4.4	234	4.3						
24, a. m ...							441	2.3	311	3.2						
25, p. m ...	179	5.6			141	7.1	147	6.8								
27, a. m ...	113	8.8			132	7.6	133	7.5								
27, p. m ...					149	6.7	146	6.8								
30, a. m ...	309	3.2			205	3.8	171	5.8	199	5.0	253	4.0	271	3.7		
31, a. m ...			149	6.7	194	5.2	244	4.1								
June 1, a. m ...					158	6.3	184	5.4								
5, a. m ...			443	2.3	122	8.2										
8, p. m ...	261	3.8	463	2.2												
11, p. m ...	192	5.2	295	3.4												
13, a. m ...			99	10.1	124	8.1	126	7.9								
18, p. m ...			273	3.7	314	3.2										
17, a. m ...			108	9.3	119	8.4	142	7.0								
17, p. m ...									146	6.8	161	6.2				
19, a. m ...	545	1.8	608	1.6			520	1.9	260	3.7	209	4.8				
20, a. m ...	796	1.2	377	2.7	538	1.9	612	1.6	234	4.3						
21, a. m ...			1,658	0.6					344	2.9	250	4.0				
22, a. m ...	132	7.6	148	6.6	185	5.4	239	4.2	394	2.5	312	3.2	225	4.4		
26, a. m ...	173	5.8	228	4.4												
27, p. m ...	106	9.3	118	8.5	146	6.8	153	6.5								
28, a. m ...	432	2.3	395	2.5												
29, a. m ...	98	10.2	116	8.6	142	7.0	157	6.4	144	6.9	170	5.9				
July 4, a. m ...	129	7.8	133	7.5	140	7.1										
5, a. m ...	158	6.3	178	5.6	196	5.1	162	6.2								
6, a. m ...	109	9.2	117	8.5	155	6.5	162	6.2								
8, p. m ...	164	6.1														
10, a. m ...	578	1.7	655	1.5	416	2.4	364	2.7	325	3.1						
11, p. m ...	140	7.1	152	6.6	180	5.6										
12, p. m ...	121	8.3														
13, a. m ...			298	3.4	355	1.2	586	1.7	314	3.2	328	3.0				
14, a. m ...	919	1.1					554	1.8	242	4.1						
15, a. m ...	156	6.4	173	5.8	160	6.2	173	5.8								
16, p. m ...	134	7.5	140	7.1	160	6.2	165	6.1								
17, a. m ...	766	1.3	495	2.0												
18, a. m ...					621	1.6	215	4.7								
19, a. m ...	332	3.0	372	2.7	366	2.7										
20, p. m	99	10.0	103	9.7	140	7.1	149	6.7	153	6.5	173	5.8				
21, a. m ...									2,399	0.4	592	1.7				
22, a. m ...	1,042	1.0														
24, a. m ...	143	7.0	162	6.2	179	5.6	244	4.1	318	4.6	242	4.1				
25, a. m ...	133	7.5	128	7.8												
26, a. m ...							1,089	0.9	626	1.6						
27, a. m ...	215	4.7					733	1.4								
28, a. m ...	124	8.1	146	6.8	170	5.9	164	6.1								
29, a. m ...	106	9.4	140	7.1	156	6.4										
Aug. 1, p. m ...	110	9.1														
2, p. m ...	106	9.4	116	8.6												
3, a. m ...	116	8.6	123	8.1	133	7.5										
4, a. m ...	357	2.8	463	2.2	1,018	1.0	565	1.8								
5, p. m ...	207	4.8														
7, a. m ...	112	8.4	128	7.8												
8, a. m ...	424	2.4	422	2.4	337	2.6										
9, a. m ...	415	2.4	410	2.4	1,939	0.5	694	1.4	353	2.8						
10, a. m ...	163	6.5	155	6.5												
13, a. m ...							365	2.7								
15, a. m ...	115	8.7			157	6.4	124	8.1								
16, a. m ...	877	1.1	5'0	1.8	1,125	0.9	1,178	0.8								
18, a. m ...	347	2.9	269	3.7												
20, a. m ...	111	9.0	118	8.4	144	6.9										
22, a. m ...	120	8.3	133	7.5												
23, a. m ...	123	8.1	123	8.1	133	7.5	227	4.4	280	3.6	235	4.3				
24, p. m ...	120	8.3	135	7.4												
26, a. m ...	127	7.9	162	6.2	237	4.2										
29, a. m ...					1,063	0.9	451	2.2	308	3.2	215	4.7				
30, a. m ...	198	5.1	436	2.3	302	1.2										
30, a. m ...			181	5.5	202	5.0										
31, a. m ...	132	7.8	128	8.1												
Sept. 1, a. m ...	477	2.1	430	2.3	266	3.8										
3, a. m ...			750	1.3	473	2.1										

Temperature—Clear—Continued.

Date.	1,000 feet. Rate.	Gradient.	1,500 feet. Rate.	Gradient.	2,000 feet. Rate.	Gradient.	3,000 feet. Rate.	Gradient.	4,000 feet. Rate.	Gradient.	5,000 feet. Rate.	Gradient.	6,000 feet. Rate.	Gradient.	Mean. Rate.	Gradient.
Sept. 4, p.m	89	11.5			115	8.7	141	7.1								
5, a.m	147	6.8	172	5.8	179	5.6										
6, a.m	934	1.1	646	1.5	486	2.1	328	3.0	237	4.2	157	6.4				
7, a.m	122	8.2	126	7.9	124	8.1	146	6.8	155	5.5						
12, a.m	164	6.1	183	5.5	178	5.6	192	5.2	235	4.3	227	4.4				
14, p.m	139	7.2														
16, a.m			188	5.3	185	5.4	190	5.3	189	5.3	191	5.3				
18, p.m	941	1.1	287	3.5	252	4.0										
19, a.m							506	1.8								
20, a.m											190	5.3				
22, n.m	88	11.4														
24, a.m	109	9.2	131	7.6	144	6.9	211	4.7								
26, p.m	77	13.0	95	10.5	112	8.9	130	7.7								
28, a.m	325	3.1	261	3.8	256	3.9	212	4.7	312	2.2						
30, a.m			655	1.5			1,017	1.0	413	2.4	272	3.7				
Oct. 1, a.m	225	4.4	213	4.7							225	4.4				
2, a.m	283	3.5	261	3.8	288	3.5										
6, a.m	165	6.1	191	5.2	257	3.9	230	4.3	292	3.4						
10, a.m	162	6.2	183	5.5	187	5.3										
13, a.m			149	6.7												
25, a.m	143	7.0	143	7.0	170	5.9										
26, a.m					447	2.2										
27, a.m	357	3.9	230	4.3	422	2.4	383	2.6	314	3.2						
28, p.m	171	5.8	153	6.5	140	7.1	175	5.7	193	5.2						
29, a.m	602	1.7	2,568	0.4	1,212	0.8	721	1.4	507	2.0						

CLOUDY.

Date.	1,000 feet. Rate.	Gradient.	1,500 feet. Rate.	Gradient.	2,000 feet. Rate.	Gradient.	3,000 feet. Rate.	Gradient.	4,000 feet. Rate.	Gradient.	5,000 feet. Rate.	Gradient.	6,000 feet. Rate.	Gradient.	Mean. Rate.	Gradient.
May 13, p.m			225	4.4	222	4.5	255	3.9								
14, p.m					153	6.5										
16, a.m			376	2.7	217	4.6										
17, p.m			316	3.2												
18, a.m			196	5.1	223	4.5										
20, a.m	195	5.1														
21, a.m			165	6.1	170	5.9										
26, a.m			340	2.9												
28, a.m			176	5.7												
28, a.m			127	7.9	138	7.2										
29, p.m	85	11.8														
June 3, a.m			352	2.4												
3, p.m					164	6.1	185	5.3								
4, a.m					331	3.0	381	2.6								
9, p.m	129	7.8	128	7.8	155	6.5										
12, a.m	457	2.2														
12, a.m	248	4.0	151	6.6												
14, a.m			239	4.2	291	3.4	184	5.4								
15, a.m			185	5.4												
22, a.m	200	5.0	151	6.6	191	5.2			1,366	0.7	409	2.4				
24, a.m	175	5.7	200	5.0	240	4.2	210	4.8	231	4.3	282	4.3				
30, a.m	1,228	0.8	910	1.1	1,020	1.0	485	2.1	306	3.2						
July 1, a.m	127	7.9	125	8.0												
2, a.m	202	5.0	195	5.1	212	4.7	269	3.7	228	4.4	273	3.7				
7, a.m					409	2.4										
9, a.m	662	1.5	944	1.1	752	1.3	427	2.3	299	3.3						
25, a.m			280	3.5	382	2.6	390	2.6	381	2.6						
30, a.m	227	4.4	252	4.0	183	7.5										
Aug. 5, p.m	160	6.2	133	7.5	239	4.2										
12, p.m	111	9.0	116	8.6	139	7.2										
14, a.m	152	6.6														
17, a.m	396	2.5	502	2.0	503	1.7	253	6.5	213	4.7						
21, a.m			1,896	0.6	596	1.7										
Sept. 8, a.m	254	3.9	290	3.4	324	3.1	344	2.9								
13, p.m	146	6.8	187	6.0	246	4.1										
29, a.m	278	3.6	306	3.3												
Oct. 2, a.m							470	2.1	446	2.2						
4, p.m	182	5.5	195	5.1												
12, a.m	318	3.1	265	3.8	291	3.4	271	3.7	381	2.6						
14, a.m							1,167	0.9								
15, a.m	220	4.5	214	4.7	286	3.5	236	4.2								
20, a.m	128	7.8	144	6.9	161	6.2	203	4.9								
21, a.m	182	5.3	185	5.4												
23, a.m	155	6.5	141	7.1	174	5.7										
24, a.m	170	5.9	192	5.2	247	4.0	260	3.7								

SUMMARY.
CLEAR.

	1,000 feet.	1,500 feet.	2,000 feet.	3,000 feet.	4,000 feet.	5,000 feet.	6,000 feet.	Mean.
Morning	5.5	5.0	4.6	4.0	3.7	4.3	4.0	4.4
Afternoon	7.4	6.5	6.5	6.4	5.7	6.0		6.4
Mean	6.0	5.3	5 0	4.5	3.9	4.5	4.0	4.7

Temperature—Continued.

CLOUDY.

Date.	1,000 feet.		1,500 feet.		2,000 feet.		3,000 feet.		4,000 feet.		5,000 feet.		6,000 feet.		Mean.	
	Rate.	Gradient.	Rate.	Gradient.	Rate.	Gradient.	Rate.	Gradient.	Rate.	Gradient.	Rate.	Gradient.	Rate.	Gradient.	Rate.	Gradient.
Morning		4.5		4.4		3.9		3.5		3.1		3.5				3.8
Afternoon		7.7		6.2		5.8		4.6								6.1
Mean		5.3		4.8		4.4		3.6		3.1		3.5				4.1

COMBINED.

Morning		5.3		4.8		4.3		3.9		3.5		4.2		4.0		4.3
Afternoon		7.5		6.4		6.2		6.1		5.7		6.0				6.3
Mean		5.9		5.1		4.8		4.3		3.7		4.4		4.0		4.6

INVERSIONS.

Date.	1,000 feet.		1,200 feet.		1,500 feet.		2,000 feet.		2,500 feet.		3,000 feet.		3,500 feet.		4,000 feet.	
	Rate.	Gradient.	Rate.	Gradient.	Rate.	Gradient.	Rate.	Gradient.	Rate.	Gradient.	Rate.	Gradient.	Rate.	Gradient.	Rate.	Gradient.
May 23			162	6.2												
24					192	5.2			2,451	0.4						
June 19							2,956	3.4								
21					1,658	0.6	1,594	0.6								
July 7			1,272	0.8												
14			910	1.1	3,932	0.3										
18	1,566	0.6														
21			250	4.0	326	3.1	994	1.0								
26	277	3.6			553	1.8	484	2.1								
27					1,530	0.7	1,304	0.8								
Aug. 11	523	1.9														
13			242	4.3	299	3.3	406	2.5								
19			173	5.8	231	4.3										
27																
Sept. 19			109	9.2	188	5.3	206	4.9	365	2.7						
20			786	1.3	104	9.6	139	7.2	222	4.5	346	2.9	585	1.7	5,014	0.2
21					1,268	0.5										
23			125	8.0	167	6.0	180	5.6	173	5.8	198	5.1	302	3.3	748	1.3
27			403	2.5	328	3.0	277	3.5								
30							1,913	0.5								
Oct. 9			215	4.7	231	4.5	290	3.4								
14			202	5.0	257	3.9	348	2.9								
26			223	4.5	232	4.3	547	1.8								
30	309	3.2			295	3.4	371	2.7								

TOPEKA, KANS.

TEMPERATURE.

The mean gradient at Topeka was 5° per thousand feet, as determined from 319 observations made during 81 ascensions. The highest altitude attained was 5,892 feet. The gradient is greatest up to 1,000 feet, and decreases steadily up to 4,000 feet, above which height it rises slowly. The decrease up to 3,000 feet is quite rapid.

The average morning gradient was 0.5° less than the mean, and decreased at about the same rate with increasing altitude. The average afternoon gradient was 1° greater than the mean, and the decrease with increase of elevation was less decided than in the morning. The clear weather gradients were greater than the cloudy ones at all elevations, but the differences were not marked except up to 2,000 and 4,000 feet, where they were 1.3° and 1.4°, respectively, per thousand feet.

Wind directions and the relative positions of high and low barometric areas did not appear to have had any effect upon the gradients. Cloud effects were confined to a suspension of the temperature fall, changing frequently to a rise, and followed by a fall as the kite descended below the clouds. There were no records of any ascents above the tops of clouds. On October 3 the temperature rose 6° while the kite was rising from 1,700 feet elevation to 2,250 feet at the clouds;

then there was a fall of 6° as the kite descended below the clouds to 1,600 feet. A second ascent to another cloud at 2,000 feet resulted in a rise of 4°, followed by a fall of 5° as the kite descended to 1,400 feet. On October 9 the temperature fell but 1° in rising from 2,900 to 4,200 feet on account of the presence of clouds.

There were several cases of temperature inversion that are worthy of notice. On May 24 there was an inversion of 11° at 2,400 feet elevation, and of 7.5° at 3,650 feet. On August 3 there was an inversion from 7.13 until 10.02 a. m. up to 3,250 feet elevation, amounting to 10° at 1,500 feet at 9.56 a. m. This inversion was due to cloud formation shortly after sunrise, and still continued after the kite had broken away at 10.05 a. m. On October 5 there was an inversion of 3° in the early morning up to 3,000 feet as the clouds cleared away, thus permitting unobstructed radiation from the surface of the earth to the lower air layers.

WINDS.

The deflections of the kite due to winds were invariably toward the right, but the general directions above and below were at all times almost exactly alike, a deflection of as much as 90° occurring but once or twice.

RELATIVE HUMIDITY AND VAPOR PRESSURE.

The relative humidity changes with increase of elevation were very small, the only differences of consequence occurring at 1,500 and 2,000 feet, where they were 8 and 5 per cent respectively, the upper air humidity being the higher. Above 2,000 feet the values are practically equal, the only differences being 1 per cent.

The approach of rain within a few hours was sometimes indicated by a rise in the upper air humidity without a corresponding change in that below, the rise usually bringing the humidity above 90 per cent. Quite frequently, however, the proximity of rain did not appear to exercise the slightest effect.

The cloud effects were not at all marked, the only noticeable one being a slight rise at times. On one day, May 2, there was a fall of 54 per cent (from 94 to 40 per cent) as the kite came near the clouds.

The vapor pressure results as given below agree very nearly with those at Omaha below 5,000 feet. Above this altitude the latter are 0.13 smaller. At 2,000, 3,000, and 4,000 feet the two are exactly alike.

DIMINUTION OF VAPOR PRESSURE WITH ALTITUDE.

VALUE OF $\frac{p}{p^o}$ AT EACH RESPECTIVE 1,000 FEET OF ALTITUDE.

1,500 feet.	2,000 feet.	3,000 feet.	4,000 feet.	5,000 feet.	6,000 feet.
0.85	0.80	0.68	0.56	0.52	0.36

TEMPERATURE.

CLEAR.

Date.	1,000 feet.		1,500 feet.		2,000 feet.		3,000 feet.		4,000 feet.		5,000 feet.		6,000 feet.		Mean.	
	Rate.	Gradient.	Rate.	Gradient.	Rate.	Gradient.	Rate.	Gradient.	Rate.	Gradient.	Rate.	Gradient.	Rate.	Gradient.	Rate.	Gradient.
May 6, a. m.					1,494	0.7	687	1.5	353	2.8	273	3.7	222	4.5		
6, p. m.									209	4.8	190	5.3				
6, p. m.							168	6.0	166	6.0						
6, a. m.					156	6.4	184	5.4								
11, a. m.	73	13.7	101	9.9	113	8.8										
11, a. m.							134	7.5	135	7.4	213	4.7				
11, a. m.											345	2.9				
11, a. m.											261	3.8				
12, a. m.			160	6.2												
14, a. m.			174	5.7												
18, a. m.	222	4.3														
23, a. m.					327	3.1	292	3.4								
23, p. m.							161	6.2	222	4.5						
24, p. m.	119	8.4	114	8.8												
29, p. m.					149	6.7	157	6.4	197	5.1						

Temperature—Clear—Continued.

Date.	1,000 feet.		1,500 feet.		2,000 feet.		3,000 feet.		4,000 feet.		5,000 feet.		6,000 feet.		Mean.	
	Rate.	Gradient.	Rate.	Gradient.	Rate.	Gradient.	Rate.	Gradient.	Rate.	Gradient.	Rate.	Gradient.	Rate.	Gradient.	Rate.	Gradient.
June 2, p. m.			149	6.7	172	5.8										
3, a. m.					672	1.5	377	2.7								
22, a. m.			1,597	0.6												
22, a. m.			536	1.9	474	2.1	300	3.3								
22, a. m.							210	4.8								
23, a. m.					781	1.3	601	1.7								
23, a. m.	111	9.0	142	7.0			233	4.3								
24, a. m.	127	7.9	164	6.1	271	3.7	217	4.6								
25, a. m.	129	7.8														
27, a. m.	241	4.1														
28, a. m.	123	8.1	144	6.9	124	8.1										
29, a. m.	139	7.2			417	2.4	339	2.9								
29, a. m.							225	4.4								
30, a. m.	192	5.2	120	8.3	199	5.0	263	3.8								
July 1, a. m.			142	7.0	643	1.6	431	2.3								
2, a. m.	144	6.9	189	5.3	171	5.8	541	1.8								
6, p. m.	164	6.1	185	5.4	276	3.6	264	3.8								
6, a. m.	154	6.5														
17, a. m.	91	11.0														
18, a. m.			163	6.1	222	4.5										
18, p. m.	118	8.5														
19, a. m.			444	2.3	147	6.8	800	1.2								
19, a. m.	131	7.6	184	7.5												
27, a. m.	106	9.4	139	7.2	141	7.1	215	4.7								
27, p. m.	72	13.9	108	9.3	140	7.1	140	7.1								
29, p. m.			180	5.6												
Aug. 4, p. m.	139	7.2			154	6.5	163	6.1								
5, p. m.	91	11.0	127	7.9	150	6.5	161	6.2								
5, p. m.	139	7.2			159	6.3										
6, a. m.	94	10.7	132	7.6	203	4.0	248	4.0								
10, p. m.	120	8.3	170	5.9	230	4.3										
14, a. m.	132	7.6	180	5.6	401	2.5	399	2.5	229	4.4	234	4.3				
15, a. m.	102	9.8			524	1.9										
15, p. m.			106	9.4												
16, a. m.	156	6.4	302	3.3	410	2.4	269	3.7								
16, p. m.	104	9.6			113	8.8										
20, p. m.			152	6.6	173	5.8	151	6.6								
20, p. m.	201	5.0			210	4.8	229	4.4								
22, a. m.	565	1.8			504	2.0	369	2.7								
22, p. m.	132	7.6	154	6.5	158	6.3	183	5.5								
23, a. m.									1,176	0.9						
23, a. m.	111	9.0	110	9.1	146	6.8	211	4.7								
28, a. m.					159	6.3	251	4.0								
28, p. m.	155	6.5	148	6.8												
29, a. m.	154	6.5	431	2.3	294	3.4										
29, p. m.			139	7.2	159	6.3										
30, a. m.	480	2.0			778	1.3	449	2.2								
30, a. m.					207	4.8										
31, a. m.					361	2.8	216	4.6								
Sept. 1, a. m.							790	1.3	551	1.8						
1, p. m.	111	9.0	114	8.8	134	7.4	228	4.4								
3, a. m.							200	5.0								
5, p. m.	99	10.1	106	9.3	135	7.4										
5, p. m.	116	8.6	150	6.7	154	6.5										
8, a. m.					399	2.5	146	6.8								
8, a. m.	139	7.2	110	9.1	150	6.7										
9, a. m.			173	5.8												
23, a. m.							1,027	1.0	217	4.6						
23, p. m.	114	8.8			155	6.5										
28, a. m.	114	8.8	185	5.4	363	3.8	247	4.0	226	4.4	202	5.0				
28, p. m.	118	8.5	152	6.6	168	6.0	171	5.8	188	5.3						
29, a. m.							805	1.2	174	5.7						
29, p. m.	110	9.1	133	7.5	138	7.2	156	6.4								
Oct. 2, a. m.							1,120	0.9								
2, p. m.	102	9.3	133	7.5	152	6.6	185	5.4								
4, p. m.	111	9.0	125	8.0												
5, a. m.	299	3.3			505	1.9	593	1.6								
5, p. m.			116	8.6												
9, a. m.	135	7.4			150	6.7	162	6.2	207	4.8						
10, p. m.	136	7.4	126	7.9												
13, a. m.	1,714	0.6			944	1.1	752	1.3	367	2.7	329	3.0				
13, p. m.	130	7.2	136	7.3	182	5.5	181	5.5	230	4.3						
14, a. m.	161	6.2	174	5.7	194	5.2	199	5.0								
14, p. m.			642	1.6	310	3.2										
15, a. m.	200	5.0					1,064	0.9	325	3.1						
29, p. m.	136	7.4	156	6.4	186	5.4	223	4.5	250	4.0	242	4.1				
29, p. m.	260	3.8					167	6.0								
30, p. m.	211	4.7	214	4.7												

Temperature—Continued.

CLOUDY.

Date.	1,000 feet. Rate.	Gradient.	1,500 feet. Rate.	Gradient.	2,000 feet. Rate.	Gradient.	3,000 feet. Rate.	Gradient	4,000 feet. Rate.	Gradient.	5,000 feet. Rate.	Gradient.	6,000 feet. Rate.	Gradient.	Mean. Rate.	Gradient.
May 2, a. m					206	4.9	333	3.0	800	1.2						
2, a. m ...									973	1.0						
10, p. m ...									221	4.5						
13, p. m ...	205	4.9														
17, a. m ...			1,638	0.6			851	1.2	575	1.7	282	4.5				
17, a. m ...									243	4.1						
19, p. m ...	171	5.8	191	5.2												
20, a. m ...					357	3.8										
21, a. m ...			215	4.7												
21, a. m ...			179	5.6												
26, a. m ...					249	4.0	230	4.3								
27, a. m ...	103	9.7			146	6.8										
31, a. m ...							1,818	0.6	793	1.3	531	1.9				
June 9, a. m	132	7.6	171	5.8	212	4.7	298	3.4								
9, p. m ...							179	5.6								
10, a. m ...			1,092	0.9			534	1.9								
16, a. m ...	102	9.8	139	7.2	187	6.4	242	4.1								
21, p. m ...			153	6.5	176	5.7	176	5.7								
26, a. m ...	112	8.9	176	5.7	242	4.1	300	3.3								
26, a. m ...							207	4.8								
July 7, a. m	239	4.2	200	5.0	241	4.1	251	4.0								
Sept. 13, a. m ...			231	4.3	205	4.9	201	5.0	203	4.3						
13, p. m ...	137	7.3			146	6.8	167	6.0								
16, a. m ...	187	5.3			314	3.2	181	5.5								
16, p. m ...			148	6.8	215	4.7										
21, p. m ...					218	4.6	204	4.9	185	5.4						
21, p. m ...	167	6.0	182	5.5	191	5.2										
30, p. m ...	131	7.6	146	6.8	185	6.1										
Oct. 3, p. m ...	125	7.4	187	5.3	753	1.3	554	1.8								
5, p. m ...	131	7.6	148	6.8	399	2.5										
9, p. m ...	129	7.8			147	6.8	162	6.2								

SUMMARY.

CLEAR.

	1,000 feet	1,500 feet	2,000 feet	3,000 feet	4,000 feet	5,000 feet	6,000 feet	Mean
Morning.....	6.9	5.8	3.9	3.4	3.9	3.9	4.5	4.6
Afternoon	8.1	7.2	6.6	5.4	4.9	4.7		6.2
Mean	7.5	6.5	5.0	4.1	4.3	4.1	4.5	5.1

CLOUDY.

	1,000 feet	1,500 feet	2,000 feet	3,000 feet	4,000 feet	5,000 feet	6,000 feet	Mean
Morning.....	7.6	4.4	4.7	3.2	2.3	3.2		4.2
Afternoon	6.8	6.1	4.9	6.0	5.0			5.6
Mean	7.1	5.2	4.8	4.0	2.9	3.2		4.5

COMBINED.

	1,000 feet	1,500 feet	2,000 feet	3,000 feet	4,000 feet	5,000 feet	6,000 feet	Mean
Morning.....	7.0	5.4	4.1	3.3	2.3	3.8	4.5	4.5
Afternoon	7.8	6.9	6.1	5.5	4.9	4.7		6.0
Mean	7.4	6.2	4.9	4.0	3.8	3.9	4.5	5.0

O

FIG. 1.—KITE WITH METEOROGRAPH IN PLACE.

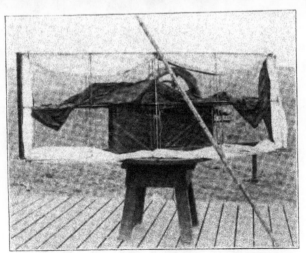

FIG. 2.—KITE COLLAPSED.

A 80° C

b

B

S'

B'

g

FIG. 3.—NORMAL BRIDLE.

—Safety line and eyes.

FIG. 4.—KITE REEL AND SUPPORT.

CPSIA information can be obtained
at www.ICGtesting.com
Printed in the USA
BVHW051847051118
532208BV00023B/4318/P